T0282649

CAMBRIDGE LIBRARY COLLECTION

Books of enduring scholarly value

Physical Sciences

From ancient times, humans have tried to understand the workings of the world around them. The roots of modern physical science go back to the very earliest mechanical devices such as levers and rollers, the mixing of paints and dyes, and the importance of the heavenly bodies in early religious observance and navigation. The physical sciences as we know them today began to emerge as independent academic subjects during the early modern period, in the work of Newton and other 'natural philosophers', and numerous sub-disciplines developed during the centuries that followed. This part of the Cambridge Library Collection is devoted to landmark publications in this area which will be of interest to historians of science concerned with individual scientists, particular discoveries, and advances in scientific method, or with the establishment and development of scientific institutions around the world.

The Works of John Playfair

John Playfair (1748–1819) was a Scottish mathematician and geologist best known for his defence of James Hutton's geological theories. He attended the University of St Andrews, completing his theological studies in 1770. In 1785 he was appointed joint Professor of Mathematics at the University of Edinburgh, and in 1805 he was elected Professor of Natural Philosophy. A Fellow of the Royal Society, he was acquainted with continental scientific developments, and was a prolific writer of scientific articles in the *Transactions of the Royal Society of Edinburgh* and the *Edinburgh Review*. This four-volume edition of his works was published in 1822 and is prefaced by a biography of Playfair. Volume 4 contains his biographies of his colleagues, and review articles on mathematical and astronomical works, both in English and French.

Cambridge University Press has long been a pioneer in the reissuing of out-of-print titles from its own backlist, producing digital reprints of books that are still sought after by scholars and students but could not be reprinted economically using traditional technology. The Cambridge Library Collection extends this activity to a wider range of books which are still of importance to researchers and professionals, either for the source material they contain, or as landmarks in the history of their academic discipline.

Drawing from the world-renowned collections in the Cambridge University Library, and guided by the advice of experts in each subject area, Cambridge University Press is using state-of-the-art scanning machines in its own Printing House to capture the content of each book selected for inclusion. The files are processed to give a consistently clear, crisp image, and the books finished to the high quality standard for which the Press is recognised around the world. The latest print-on-demand technology ensures that the books will remain available indefinitely, and that orders for single or multiple copies can quickly be supplied.

The Cambridge Library Collection will bring back to life books of enduring scholarly value (including out-of-copyright works originally issued by other publishers) across a wide range of disciplines in the humanities and social sciences and in science and technology.

The Works of John Playfair

VOLUME 4

JOHN PLAYFAIR
EDITED BY JAMES G. PLAYFAIR

CAMBRIDGE
UNIVERSITY PRESS

CAMBRIDGE UNIVERSITY PRESS

Cambridge, New York, Melbourne, Madrid, Cape Town,
Singapore, São Paolo, Delhi, Tokyo, Mexico City

Published in the United States of America by Cambridge University Press, New York

www.cambridge.org
Information on this title: www.cambridge.org/9781108029414

© in this compilation Cambridge University Press 2011

This edition first published 1822
This digitally printed version 2011

ISBN 978-1-108-02941-4 Paperback

ILLUSTRATIONS

OF THE

HUTTONIAN THEORY.

" *Nunc naturalem causam quærimus et assiduam, non raram et fortuitam.*"

SENECA.

BIOGRAPHICAL ACCOUNT

OF THE LATE

MATTHEW STEWART, D.D. F.R.S. Edin,

AND PROFESSOR OF MATHEMATICS IN THE

UNIVERSITY OF EDINBURGH.

———————

BIOGRAPHICAL ACCOUNT

OF

MATTHEW STEWART, D. D. *

The Reverend Dr Matthew Stewart, late Professor of Mathematics in the University of Edinburgh,
was the son of the Reverend Mr Dugald Stewart,
Minister of Rothsay in the Isle of Bute, and was
born at that place in the year 1717. After having
finished his course at the grammar-school, being
intended by his father for the church, he was sent
to the University of Glasgow, and was entered
there as a student in 1734. His academical studies
were prosecuted with diligence and success; and he
was so happy as to be particularly distinguished by
the friendship of Dr Hutcheson and Dr Simson.
With the latter, indeed, he soon became very intimately connected; for though it is said, that his
predilection for the mathematics did not instantly
appear on his application to the study of that
science, yet the particular direction of his talents
was probably observed by his master before it was

---

* From the Transactions of the Royal Society of Edinburgh, Vol. I. (1788.)—Ed.

perceived by himself. Accordingly, after being
the pupil of Dr Simson, he became his friend;
and during all the time that he remained at the
University of Glasgow, pursuing the studies of philosophy and theology, he lived in the closest intimacy with that excellent mathematician, and was
instructed by him in, what might not improperly be
called, the *arcana* of the ancient geometry. That
science was yet involved in some degree of mystery;
for though the extent of its discoveries was nearly
ascertained, its analysis, or method of investigation,
was but imperfectly understood, and seemed inadequate to the discoveries which had been made by it.
The learning and genius of Viviani, Fermat, Halley, and of other excellent mathematicians, had already been employed in removing this difficulty;
but their efforts had not been attended with complete success. Dr Simson was now engaged in
perfecting what they had begun, and in resisting
the encroachments, which he conceived the modern analysis to be making upon the ancient. With
this view, he had already published a treatise of
Conic Sections, and was now preparing a restoration of the Loci Plani of Apollonius, in which that
work was to resume its original elegance and simplicity. To these, and other studies of the same
kind, he constantly directed the attention of his
young friend, while he was delighted, and astonished at the rapidity of his progress.

Mr Stewart's views made it necessary for him to attend the lectures in the University of Edinburgh in 1741 ; and that his mathematical studies might suffer no interruption, he was introduced by Dr Simson to Mr Maclaurin, who was then teaching, with so much success, both the geometry and the philosophy of Newton. Mr Stewart attended his lectures, and made that proficiency which was to be expected from the abilities of such a pupil, directed by those of so great a master. But the modern analysis, even when thus powerfully recommended, was not able to withdraw his attention from the ancient geometry. He kept up a regular correspondence with Dr Simson, giving him an account of his progress, and of his discoveries in geometry, which were now both numerous and important, and receiving in return many curious communications with respect to the *Loci Plani*, and the Porisms of Euclid. These last formed the most intricate and paradoxical subject in the history of the ancient mathematics. Every thing concerning them, but the name, had perished. Pappus Alexandrinus has made mention of three books of Porisms written by Euclid, and has given an account of what they contained ; but this account has suffered so much from the injuries of time, that the sense of one proposition only is complete. There was no diagram to direct the geometer in his researches, nor any general notion of the subject, or of the form of the propositions, to serve as a rule for his conjectures. The task, therefore, of restoring these ancient books, which Dr Simson now imposed on himself, exceeded infinitely the ordinary labours of the critic or the antiquary ; and it was only by uniting the learning and diligence of these two characters, with the skill of a profound geometer, that he was at last successful in this difficult undertaking. He had begun it as early as the year 1727, but seems to have communicated the whole progress of his discoveries to Mr Stewart alone.

While the second invention of Porisms, to which more genius was perhaps required than to the first discovery of them, employed Dr Simson, Mr Stewart pursued the same subject in a different, and new direction. In doing so, he was led to the discovery of those curious and interesting propositions, which were published, under the title of General Theorems, in 1746. They were given without the demonstrations ; but did not fail to place their discoverer at once among the geometers of the first rank. They are, for the most part, Porisms, though Mr Stewart, careful not to anticipate the discoveries of his friend, gave them no other name than that of Theorems. They are among the most beautiful, as well as most general propositions known in the whole compass of geometry, and are perhaps only equalled by the remarkable *Locus* to the circle in the second book of Apollonius, or by the celebrated theorem of Mr Cotes. The first demonstration of any considerable number of them, is that which was lately communicated to this Society, * though I believe there are few mathematicians, into whose hands they have fallen, whose skill they have not often exercised. The unity which prevails among them is a proof, that a single, though extensive view, guided Mr Stewart in the discovery of them all. It seems probable, that, while he aimed at extending geometry beyond the limits it had reached with the ancients, he had begun to consider the *Locus ad quatuor rectas*, beyond which their analysis had not reached. With this view, he, no doubt, thought of extending the hypotheses of that problem to their utmost generality ; that is, to any number of perpendiculars drawn to an equal number of lines, and to any power whatever of these perpendiculars. In doing this, he could not fail to meet with many curious porisms ; for a porism is nothing else than that particular case, when the data of a problem are so related to one another, as to render it indefinite, or capable of innumerable solutions. These cases, which rarely occur, except in the construction of very general and complicated problems, must always interest a geometer, because they trace out the di-

---

* By the Reverend Dr Small.

visions of his subject, and are usually distinguished
by an elegance and simplicity peculiar to themselves. Such, accordingly, were the propositions
which Mr Stewart now communicated to the world.
He suppressed his investigations, however, which
were geometrical, and which, if given with all the
precision required by the forms of the ancient geometry, would probably have occupied several volumes.

The history of these geometrical discoveries has
led us to neglect the order of time. For Mr
Stewart, while engaged in them, had entered into
the church, and, through the patronage of the Earl
of Bute and the Duke of Argyle, had obtained the
living of Roseneath. It was in that retired and
romantic situation, that he discovered the greater
part of the propositions that have just been mentioned. There, also, he used to receive the visits
of his friend Mr Melvil, whose ingenious observations in the *Physical and Literary Essays* give
us cause to regret that he was so early taken from
the world of science. *

In the summer of 1746, the mathematical chair
in the University of Edinburgh became vacant by
the death of Mr Maclaurin. The General Theorems had not yet appeared ; Mr Stewart was known
only to his friends ; and the eyes of the public were

* Observations on Light and Colours, Phys. and Lit.
Fssays, Vol. II. Art. 4.

naturally turned on Mr Stirling, who then resided at Leadhills, and who was well known in the mathematical world. He, however, declined appearing as a candidate for the vacant chair ; and several others were named, among whom was Mr Stewart. In the end of this year, the General Theorems were published, and gave to their author a decided superiority above all the other candidates. He was accordingly elected Professor of Mathematics in the University of Edinburgh, in the beginning of September 1747.

The duties of this office gave a turn somewhat different to his mathematical pursuits, and led him to think of the most simple and elegant means of explaining those difficult propositions, which were hitherto only accessible to men deeply versed in the modern analysis. In doing this, he was pursuing the object which, of all others, he most ardently wished to attain, viz. the application of geometry to such problems as the algebraic calculus alone had been thought able to resolve. His solution of Kepler's problem was the first specimen of this kind which he gave to the world ; and it was impossible to have produced one more to the credit of the method he followed, or of the abilities with which he applied it. When the astronomer, from whom that problem takes its name, discovered the elliptical motion of the planets, and their equable description of areas round the sun, he reduced the problem, of computing the place of a planet for a given time, to that of drawing a line through the focus of an ellipse, that should divide the area of the semi-ellipse in a given ratio. It was soon found, that this problem did not admit of an accurate solution ; and that no more was to be expected, than an easy and exact approximation. In this, ever since the days of Kepler, the mathematicians of the first name had been engaged, and the utmost resources of the integral calculus had been employed. But though many excellent solutions had been given, there was none of them at once direct in its method and simple in its principles. Mr Stewart was so happy as to attain both these objects. He founds his solution on a general property of curves, which, though very simple, had perhaps never been observed ; and, by a most ingenious application of that property, he shows how the approximation may be continued to any degree of accuracy, in a series of results which converge with prodigious rapidity. Whoever examines this solution will be astonished to find a problem brought down to the level of elementary geometry, which had hitherto seemed to require the finding of fluents and the reversion of series ; he will acknowledge the reasonableness of whatever confidence Mr Stewart may be hereafter found to place in those simple methods of investigation, which he could conduct with so much ingenuity and success ; and will be convinced, that the solution of a problem, though the most elementary, may be the least ob‑vious, and, though the easiest to be understood, may be the most difficult to be discovered.

This solution appeared in the second volume of the Essays of the Philosophical Society of Edinburgh, for the year 1756. In the first volume of the same collection, there are some other propositions of Mr Stewart's, which are an extension of a curious theorem in the fourth book of Pappus. They have a relation to the subject of porisms, and one of them forms the 91st of Dr Simson's Restoration. They are besides very beautiful propositions, and are demonstrated with all the elegance and simplicity of the ancient analysis.

It has been already mentioned, that Mr Stewart had formed the plan of introducing into the higher parts of mixed mathematics the strict and simple form of ancient demonstration. The prosecution of this plan produced the Tracts Physical and Mathematical, which were published in 1761. In the first of these, Mr Stewart lays down the doctrine of centripetal forces, in a series of propositions, demonstrated (if we admit the quadrature of curves) with the utmost rigour, and requiring no previous knowledge of the mathematics, except the elements of plane geometry, and of conic sections. The good order of these propositions, added to the clearness and simplicity of the demonstrations,

renders this tract the best elementary treatise
of physical astronomy that is any where to be
found.

In the three remaining tracts, our author had it
in view to determine, by the same rigorous method,
the effect of those forces which disturb the motions
of a secondary planet.   From this he proposed to
deduce, not only a theory of the moon, but a determination of the sun's distance from the earth.
The former is well known to be the most difficult
subject to which mathematics have been applied.
Though begun by Sir Isaac Newton, and explained, as to its principles, with singular success; yet,
as to the full detail and particular explanation
of each irregularity, it was left by that great philosopher less perfect than any other of his researches.
Succeeding mathematicians had been employed
about the same subject; the problem of the Three
Bodies had been proposed in all its generality, and,
in as far as regards the motion of the moon, had
been resolved by a direct and accurate approximation.   But the intricacy and length of these calculations rendered them intelligible only to those,
who were well versed in the higher parts of the mathematics.   This was what Dr Stewart proposed to
remedy, by giving a theory of the moon that might
depend, if possible, on elementary geometry alone,
or which should, at least, be the simplest that the
nature of things would allow.   The Tracts were

destined to serve as the basis of this investigation. We are not, however, to imagine, that Dr Stewart intended to proceed in the same direct manner that Clairault, and some other geometers, had done. It is not probable that he believed this to be within the power of pure geometry. His design undoubtedly was, to pursue that method of approximation which Sir Isaac Newton had begun, and which Callendrini, Machin, and Walmsley, had greatly improved; and, by using the methods of geometry, he hoped to reduce the problem to its ultimate simplicity. Such an undertaking was worthy of a great geometer, and of a philosopher, who considered that one of the chief obstructions to the advancement of knowledge, is the difficulty of simplifying that knowledge, which has already been acquired. We must regret, therefore, that the decline of Dr Stewart's health, which began soon after the publication of the tracts, did not permit him to pursue this investigation.

The other object of the Tracts was to determine the distance of the sun, from his effect in disturbing the motions of the moon. The approach of the transit of Venus, which was to happen in 1761, had turned the attention of mathematicians to the solution of this curious problem. But when it was considered, of how delicate a nature the observations were from which that solution was to be deduced, and to how many accidents they were exposed, it was natural, that some attempt should be made to ascertain the dimensions of our system, by means less subject to disappointment. Such accordingly was the design of Dr Stewart; and his inquiries into the lunar irregularities had furnished him with the means of accomplishing it.

The theory of the composition and resolution of forces enables us to determine what part of the solar force is employed in disturbing the motions of the moon; and, therefore, could we measure the instantaneous effect of that force, or the number of feet by which it accelerates or retards the moon's motion in a second, we should be able to determine how many feet the whole force of the sun would make a body, at the distance of the moon, or of the earth, descend in a second, and, consequently, how much the earth is, in every instant, turned out of its rectilineal course. Thus, the curvature of the earth's orbit, or, which is the same thing, the radius of that orbit, that is, the distance of the sun from the earth, would be determined. But the fact is, that the instantaneous effects of the sun's disturbing force are too minute to be measured; and that it is only the effect of that force, continued for an entire revolution, or some considerable portion of a revolution, which astronomers are able to observe.

There is yet a greater difficulty which embarrasses the solution of this problem. For, as it is only

12

by the difference of the forces exerted by the sun on
the earth and on the moon, that the motions of the
latter are disturbed, the farther off the sun is supposed, the less will be the force by which he disturbs
the moon's motions; yet that force will not diminish beyond a fixed limit, and a certain disturbance
would obtain, even if the distance of the sun were
infinite. Now the sun is actually placed at so
great a distance, that all the disturbances, which he
produces on the lunar motions, are very near to this
limit, and, therefore, a small mistake in estimating
their quantity, or in reasoning about them, may
give the distance of the sun infinite, or even impossible. But all this did not deter Dr Stewart
from undertaking the solution of the problem, with
no other assistance than that which geometry could
afford. Indeed, the idea of such a problem had
first occurred to Mr Machin, who, in his book on
the laws of the moon's motion, has just mentioned
it, and given the result of a rude calculation, (the
method of which he does not explain,) which
assigns 8$''$ for the parallax of the sun. He made
use of the motion of the nodes, but Dr Stewart
considered the motion of the apogee, or of the
longer axis of the moon's orbit, as the irregularity
best adapted to his purpose. It is well known,
that the orbit of the moon is not immoveable, but
that, in consequence of the disturbing force of the
sun, the longer axis of that orbit has an angular

motion, by which it goes back about three degrees in every lunation, and completes an entire revolution in nine years nearly. This motion, though very remarkable and easily determined, has the same fault, in respect of the present problem, that was ascribed to the other irregularities of the moon; for a very small part of it only depends on the parallax of the sun; and of this Dr Stewart, as will afterwards appear, seems not to have been perfectly aware.

The propositions, however, which defined the relation between the sun's distance and the mean motion of the apogee, were published among the Tracts in 1761. The transit of Venus happened in that same year : the astronomers returned, who had viewed that curious phenomenon from the most distant stations; and no very satisfactory result was obtained from a comparison of their observations. Dr Stewart then resolved to apply the principles he had already laid down ; and, in 1763, he published his essay on the sun's distance, where the computation being actually made, the parallax of the sun was found to be no more than 6″.9. and his distance, of consequence, almost 29875 semidiameters of the earth. *

A determination of the sun's distance, that so far exceeded all former estimations of it, was re-

---

* About 118,541,428 English miles.

ceived with surprise, and the reasoning on which it
was founded was likely to be subjected to a severe
examination. But, even among astronomers, it
was not every one who could judge in a matter of
such difficult discussion. Accordingly, it was not
till about five years after the publication of the
Sun's Distance, that there appeared a pamphlet,
under the title of Four Propositions, intended to
point out certain errors in Dr Stewart's investigation, which had given a result much greater than
the truth. A dispute in geometry was matter of
wonder to many, and perhaps of satisfaction to some,
who envied that science the certainty of its conclusions. On account of such, it must be observed,
that there are problems so extremely difficult, that,
in the solution of them, it is possible only to approximate to the truth ; and that, as in arithmetic,
we neglect those small fractions, which, though of
inconsiderable amount, would exceedingly embarrass our computations ; so, in geometry, it is sometimes necessary to reject those small quantities,
which would add little to the accuracy, and much
to the difficulty of the investigation. In both cases,
however, the same thing may happen ; though each
quantity thrown out may be inconsiderable in itself,
yet the amount of them altogether, and their effect
on the last result, may be greater than is appre-
hended. This was just what had happened in the
present case. The problem to be resolved is, in its

nature, so complex, and involves the estimation of
so many causes, that, to avoid inextricable difficulties, it is necessary to reject some quantities, as
being small in comparison of the rest, and to reason
as if they had no existence. Dr Stewart, too, it
must be confessed, had an additional motive for
wishing to simplify his investigation.   This was,
his resolution to employ in it no other method than
the geometrical, which, however excellent in other
respects, is inferior to algebra, for the conducting
of very complicated reasonings.   The skill of this
most profound and experienced geometer could not
remedy that defect ; and he was reduced to the
necessity of rejecting quantities, which were considerable enough to have a great effect on the last result.   An error was thereby introduced, which, had
it not been for certain compensations, would have
become immediately obvious, by giving the sun's
distance near three times as great as that which has
been mentioned.

   The author of the pamphlet, referred to above,
was the first who remarked the dangerous nature of
these simplifications, and who attempted to estimate
the error to which they had given rise.   In this
last, however, he has not completely succeeded ;
and that, too, by committing a mistake similar to
that which he censured in Dr Stewart, and by rejecting quantities not less than some which he retained.   He observed, however, what produced the

compensation that has been taken notice of, viz. the immense variation of the sun's distance, which corresponds to a very small variation of the motion of the moon's apogee. It is doubtful, whether Dr Stewart was fully apprized of this circumstance ; because the geometrical method, elegant and beautiful as it is, rarely presents a general view of the relations, which the magnitudes it treats of bear to one another ; and many of these relations may, therefore, escape the most profound geometer, which an algebraist, of more ordinary abilities, would not have failed to discover.

There are other of this author's strictures which we cannot admit as just, but which we will not attempt here either to enumerate or refute. Yet it were doing great injustice to his remarks, not to acknowledge, that, besides being just in the points already mentioned, they are, every where, ingenious, and written with much modesty and good temper. The author, who concealed his name, and permits it now, for the first time, to be publicly mentioned, was Mr Dawson, a surgeon at Sudbury in Yorkshire ; a man, as it should seem, who might have enjoyed more of the fame, had he been less satisfied with the possession of knowledge.

A second attack was soon after this made on the Sun's Distance, by Mr Landen ; but by no means with the same good temper which has been remarked in the former. He fancied to himself errors in

Dr Stewart's investigation, which have no existence ; he exaggerated those that were real, and
seemed to triumph in the discovery of them with
unbecoming exultation. If there are any subjects
on which men may be expected to reason dispas-
sionately, they are certainly the properties of number and extension ; and whatever pretexts moralists
or divines may have for abusing one another, mathematicians can lay claim to no such indulgence.
The asperity of Mr Landen's animadversions must
not, therefore, pass uncensured, though it be united with sound reasoning and accurate discussion.
The error into which Dr Stewart had fallen, though
before taken notice of by Mr Dawson, was first exactly determined in the work before us. * But Mr
Landen, in the zeal of correction, brings many
other charges against Dr Stewart, the greater part
of which seem to have no good foundation. Such
are his objections to the second part of the investigation, where Dr Stewart finds the relation between
the disturbing force of the sun, and the motion of
the apsides of the lunar orbit. For this part, instead of being liable to objection, is deserving of the

---

* It is but justice to remark, that Mr Landen had probably never seen Mr Dawson's Propositions at the time his own
were published, the whole impression of them, almost, having been burnt by a fire which consumed the warehouse
where they were lodged.

greatest praise, since it resolves, by geometry alone,
a problem which had eluded the efforts of some of
the ablest mathematicians, even when they availed
themselves of the utmost resources of the integral
calculus. Sir Isaac Newton, though he assumed
the disturbing force very near the truth, computed
the motion of the apsides from thence only at one
half of what it amounts to in reality ; and so, had
he been required, like Dr Stewart, to invert the
problem, he would have committed an error, not
merely of a few thousandth parts, as the latter is
alleged to have done, but would have brought out
a result double of the truth. * Machin and Callendrini, when commenting on this part of the Principia, found a like inconsistency between their theory and observation. Three other celebrated mathematicians, Clairault, D'Alembert, and Euler,
separately experienced the same difficulties, and
were led into an error of the same magnitude. It
is true, that, on resuming their computations, they
found, that they had not carried their approximations to a sufficient length, which, when they had
at last accomplished, their results agreed exactly
with observation. Mr Walmsley and Dr Stewart
were, I think, the first mathematicians who, employing in the solution of this difficult problem, the
one the algebraic calculus, and the other the geo-

* Prin. Math. Lib. 3. Prop. 3.

metrical method, were led immediately to the truth ;
a circumstance so much for the honour of both, that
it ought, by no means, to be forgotten. It was
the business of an impartial critic, while he examined our author's reasonings, to have remarked, and
to have weighed these considerations.

We may add, that the accurate measurement of
the sun's distance, and the complete theory of the
moon's motions, with which science has been en-
riched, since the time to which we now refer, sufficiently vindicate the principle of Dr Stewart's investigation, and show how much reason he had to
expect, that the former might be inferred from the
latter with considerable exactness. M. Mayer,
from one of the lunar irregularities, computes the
sun's parallax to be 7″.8, nearly a mean between
the parallax already mentioned, and that which has
been deduced from the transit of Venus in 1769. *

On the whole, therefore, while it must be acknowledged, that Dr Stewart's determination of
the sun's distance is, by no means, free from error,
it may safely be asserted, that it contains a great
deal which will always interest geometers, and always be admired by them. Few errors in science
are redeemed by the display of so much ingenuity,
and what is more singular, of so much sound reasoning. The investigation is every where elegant,

---

* Theoria Lunæ, Sect. 51.

and will, probably, be long regarded as a specimen
of the most arduous inquiry which has been attempted by mere geometry ; at the same time, the
mistake into which the geometrical method has be-
trayed this great mathematician, will serve as a
proof that it is not equal to such difficult researches;
and that in those cases, especially, where approximation is to be used, it is necessary to sacrifice the
rigour of the ancient demonstration for the accuracy of the modern analysis.

The Sun's Distance was the last work which Dr
Stewart published ; and though he lived to see the
animadversions made on it, that have been taken
notice of above, he declined entering into any controversy.   His disposition was far from polemical ;
and he knew the value of that quiet, which a literary man should rarely suffer his antagonists to in-
terrupt.   He used to say, that the decision of the
point in question was now before the public ; that,
if his investigation was right, it would never be
overturned, and that, if it was wrong, it ought not
to be defended.

A few months before he published the Essay just
mentioned, he gave to the world another work, entitled, *Propositiones Geometricæ More Veterum
Demonstratæ.*   This title, I have been told, was
given it by Dr Simson, who rejoiced in the publication of a work so well calculated to promote the
study of the ancient geometry.   It consists of a se-

ries of geometrical theorems, for the most part
new ; investigated, first, by an analysis, and afterwards synthetically demonstrated by the inversion
of the same analysis. In the former, the proposition to be investigated is supposed true ; from
thence consequences are deduced, and the reasoning is carried on till some consequence is drawn
that is already known to be true. A necessary connection is thus traced between the proposition that
was supposed true, and another that is certainly
known to be so ; and, thus, an ingenious method
is laid down for making the knowledge of any truth
subservient to the discovery of its demonstration.
This method made an important part in the analysis
of the ancient Geometers ; but few examples of it
have been preserved in their writings, and those in
the *Propositiones Geometricæ* are, on that account,
the more valuable.

Dr Stewart's constant use of the geometrical
analysis had put him in possession of many valuable propositions, which did not enter into the plan
of any of the works that have been enumerated.
Of these, not a few have found a place in the writings of Dr Simson, where they will for ever remain,
to mark the friendship of these two Mathematicians,
and to evince the esteem which Dr Simson entertained for the abilities of his pupil. In the preface
to his Conic Sections, in which he acknowledges,
that all the theorems, distinguished by the letter

x, were communications from Dr Stewart, he calls
him, " egregiæ indolis et peritiæ virum ;" and in
that to his Porisms, after pointing out many propositions that had been suggested by Pappus, and a
few that had been adopted from Fermat, he adds,
" Alia quædam adjecta sunt quorum præcipua mihi
proposuit, et aliquorum constructionem dedit exi-
mius Geometra Matthæus Stewart, a quo materia
hæc jam egregie est exculta, postea, ut spero, mul-
tum excolenda."

There is also a theorem of Dr Stewart's published in Dr Simson's edition of Euclid's *Data*, which
I take notice of, chiefly as it affords me an opportunity of paying a tribute to the memory of a man,
whose high rank did not prevent him from cultivat-
ing a science, which it enabled him to patronize.
In the note, where Dr Simson acknowledges that
communication, he mentions another theorem, also
published among the *Data;* " These propositions
(says he) were communicated to me by two excellent Geometers, the first by the Earl Stanhope, the
second by Dr Matthew Stewart."

To this Nobleman, for whose abilities and worth
Dr Stewart entertained the highest respect, he
made a visit in the course of a tour through England, soon after the publication of the Essay on the
Sun's Distance, and received from him very singular marks of attention. At a later period, when
he lamented the loss of Dr Simson, he had the con-

solation to see a lasting monument raised to the
fame of his friend, by the munificence of Lord
Stanhope, who, by the publication of Dr Simson's
posthumous works, has obliged the world with a
restoration of the most curious fragment of the
Greek geometry.

Soon after the publication of the Sun's Distance,
Dr Stewart's health began to decline, and the duties of his office became burdensome to him.   In
the year 1772, he retired to the country, where he
afterwards spent the greater part of his life, and
never resumed his labours in the University.   He
was, however, so fortunate as to have a son, to
whom, though very young, he could commit the
care of them with the greatest confidence.   Mr
Dugald Stewart, having begun to give lectures for
his father from the period above mentioned, was
elected joint Professor with him in 1775, and gave
an early specimen of those abilities, which have not
been confined to a single science.

After mathematical studies (on account of the
bad state of health into which Dr Stewart was now
falling) had ceased to be his business, they continued to be his amusement.   The analogy between
the circle and hyperbola had been an early object
of his admiration.   The extensive views which that
analogy is continually opening ; the alternate appearance and disappearance of resemblance in the
midst of so much dissimilitude, make it an object

that astonishes the experienced as well as the young
geometer.   To the consideration of this analogy,
therefore, the mind of Dr Stewart very naturally
returned, when disengaged from other speculations.
His usual success still attended his investigations ;
and he has left, among his papers, some curious
approximations to the areas, both of the circle and
hyperbola.   For some years toward the end of his
life, his health scarcely allowed him to prosecute
study even as an amusement.   He died January
23, 1785, at the age of 68.

The habits of study, in a man of original genius,
are objects of curiosity, and deserve to be remembered.   Concerning those of Dr Stewart, his writings have made it unnecessary to remark, that, from
his youth, he had been accustomed to the most intense and continued application.   In consequence
of this application, added to the natural vigour of
his mind, he retained the memory of his discoveries
in a manner that will hardly be believed.   He rarely wrote down any of his investigations, till it became necessary to do so for the purpose of publication.   When he discovered any proposition, he
would put down the enunciation with great accuracy, and, on the same piece of paper, would con-
struct very neatly the figure to which it referred.
To these he trusted for recalling to his mind, at
any future period, the demonstration or the analysis, however complicated it might be.   Experience

had taught him, that he might place this confidence
in himself without any danger of disappointment ;
and for this singular power, he was probably more
indebted to the activity of his invention, than the
mere tenaciousness of his memory.

Though he was extremely studious, he read few
books, and verified the observation of M. D'Alembert, that, of all the men of letters, mathematicians
read least of the writings of one another.   His own
investigations occupied him sufficiently ; and, indeed, the world would have had reason to regret
the misapplication of his talents, had he employed,
in the mere acquisition of knowledge, that time
which he could dedicate to works of invention.

It was his custom to spend the summer at a de-
lightful retreat in Ayrshire, where, after the academical labours of the winter were ended, he found
the leisure necessary for the prosecution of his researches.   In his way thither, he frequently made
a visit to Dr Simson at Glasgow, with whom he
had lived from his youth in the most cordial and
uninterrupted friendship.   It was pleasing to observe, in these two profound mathematicians, the
most perfect esteem and affection for each other,
and the most entire absence of jealousy, though no
two men ever trode more nearly in the same path.
The similitude of their pursuits, as it will ever do
with men superior to envy, served only to endear
them to one another.   Their sentiments and views

of the science they cultivated were nearly the same ; they were both profound geometers ; they equally admired the ancient mathematicians, and were equally versed in their methods of investigation; and they were both apprehensive that the beauty of their favourite science would be forgotten for the less elegant methods of algebraic computation. * This innovation they endeavoured to oppose ; the one, by reviving those books of the ancient geometry which were lost; the other, by extending that geometry to the most difficult inquiries of the moderns. Dr Stewart, in particular, had remarked the intricacies in which many of the greatest of the modern mathematicians had involved themselves in the application of the calculus, which a little attention to the ancient geometry would certainly have enabled them to avoid. He had observed, too, the elegant synthetical demonstrations that, on many occasions, may be given of the most difficult propositions, investigated by the inverse method of

* On the reverse of a miniature picture of Dr Simson, now in the possession of Mr Professor Stewart, is an inscription written by Dr Moore, late Professor of Greek at Glasgow, an intimate friend of Dr Simson, and a great admirer of the ancient geometry :

GEOMETRIAM, SUB TYRANNO BARBARO SAEVA SERVITUTE DIU SQUALENTEM, IN LIBERTATEM ET DECUS ANTIQUUM VINDICAVIT UNUS.

fluxions.   These circumstances had, perhaps, made
a stronger impression than they ought, on a mind
already filled with admiration of the ancient geometry, and produced too unfavourable an opinion
of the modern analysis.   But, if it be confessed
that Dr Stewart rated, in any respect too high, the
merit of the former of these sciences, this may well
be excused in the man whom it had conducted to
the discovery of the *General Theorems,* to the *solution of* Kepler's *Problem,* and to an *accurate* determination of the *sun's disturbing force.*   His
great modesty made him ascribe to the method he
used that success which he owed to his own abilities.

FINIS.

BIOGRAPHICAL ACCOUNT

OF THE LATE

JAMES HUTTON, M.D. F.R.S. Edin.

BIOGRAPHICAL ACCOUNT

OF

JAMES HUTTON, M. D. *

Dʀ Jᴀᴍᴇs Hᴜᴛᴛᴏɴ was the son of Mr William
Hutton, merchant in Edinburgh, and was born in
that city on the 3d of June 1726.   His father, a
man highly respected for his good sense and integri-
ty, and who for some years held the office of City
Treasurer, died while James was very young.   The
care of her son's education devolved, of course, on
Mrs Hutton, who appears to have been well qualified for discharging this double portion of parental
duty.   She resolved to bestow on him a liberal
education, and sent him first to the High School of
Edinburgh, and afterwards to the University, where
he entered as a student of humanity in November
1740.
  Of the masters under whom he studied there,
Maclaurin was by far the most eminent; and Dr
Hutton, though he had cultivated the mathematical sciences less than any other, never mentioned

---

* From the Transactions of the Royal Society of Edinburgh, Vol. V. (1805.)—Eᴅ.

the lectures of that celebrated professor but in
terms of high admiration.

He used also to acknowledge his obligations to
Professor Stevenson's Prelections on Logic ; not so
much, however, for having made him a logician as
a chemist.   The fact that gold is dissolved in *aqua
regia*, and that two acids which can each of them
singly dissolve any of the baser metals, must unite
their strength before they can attack the most pre-
cious, was mentioned by the professor as an illustration of some general doctrine.   The instinct of genius, if I may call it so, enabled Mr Hutton, young
as he then was, to feel, probably, rather than to understand, the importance of this phenomenon ;
and as if, by the original constitution of his mind,
a kind of elective attraction had drawn him towards chemistry, he became from that moment attached to it by a force that could never afterwards
be overcome.   He made an immediate search for
books that might give him some farther instruction
concerning the fact which he had just heard of ; but
the only one he could procure, for a long time, was
Harris's Lexicon Technicum, the predecessor of
those voluminous compilations which have since
contributed so much more to extend the surface,
than to increase the solidity of science.   It was
from the imperfect sketch contained in that dictionary, that he derived his first knowledge of chemistry, his love for which never forsook him after-

wards, and was in truth the propensity which decided the whole course and complexion of his future life.

Though his taste and capacity for instruction
were sufficiently conspicuous during his course
of academical study, his friends wished him rather
to pursue business than science. This was a measure by no means congenial to his mind, yet he acquiesced in it without difficulty.

Accordingly, in 1743 he was placed as an apprentice with Mr George Chalmers, writer to the
signet ; and subjection to the routine of a laborious
employment, was now about to check the ardour
and repress the originality of a mind formed for
different pursuits. But happily the force of genius cannot always be controlled by the plans of a
narrow and short-sighted prudence. The young
man's propensity to study continued, and he was
often found amusing himself and his fellow appren-
tices with chemical experiments, when he should
have been copying papers, or studying the forms
of legal proceedings ; so that Mr Chalmers soon
perceived that the business of a writer was not that
in which he was destined to succeed. With much
good sense and kindness, therefore, he advised him
to think of some employment better suited to his
turn of mind, and released him from the obligations
which he had come under as his apprentice. In
this he did an essential service to science, and to

the young man himself. A man of talents may follow any profession with advantage ; a man of genius will hardly succeed but in that which nature has pointed out.

The study of medicine, as being the most nearly allied to chemistry, was that to which young Hutton now resolved to dedicate his time.   He began that study under Dr George Young, the father of the late Dr Thomas Young, and at the same time attended the lectures in the university.   This course of medical instruction he followed from 1744 to 1747.

Though a regular school of medicine had now been established in the University of Edinburgh for several years, the system of medical education was neither in reality, nor in the opinion of the world, so complete as it has since become.   Some part of a physician's studies was still to be prosecuted on the Continent ; and, accordingly, in the end of 1747, Mr Hutton repaired to Paris, where he pursued with great ardour the studies of chemistry and anatomy.   After remaining in that metropolis nearly two years, he returned by the way of the Low Countries, and took the degree of Doctor of Medicine at Leyden in September 1749.   His thesis is entitled, De Sanguine et Circulatione in Microcosmo.

On his return to London about the end of that year, he began to think seriously of settling in the world. His native city, to which his views of course were first turned, afforded no very flattering prospect for his establishment as a physician. The business there was in the hands of a few eminent practitioners who had been long established ; so that no opening was left for a young man whose merit was yet unknown, who had no powerful connections to assist him on his first outset, and very little of that patient and circumspect activity by which a man pushes himself forward in the world.

These considerations seem to have made a very deep impression on his mind, and he wrote on the subject of his future prospects with considerable anxiety to his friends in Edinburgh.

One of these friends was Mr James Davie, a young man nearly of his own age, with whom he had early contracted a very intimate friendship, that endured through the whole of his life, without interruption, to the mutual benefit of both. The turn which both of them had for chemical experiments formed their first connection, and cemented it afterwards. They had begun together to make experiments on the nature and production of sal ammoniac. These experiments had led to some valuable discoveries, and had been farther pursued by Mr Davie during Dr Hutton's absence. The result afforded a reasonable expectation of establishing a profitable manufacture of the salt just named from coal-soot.

The project of this establishment was communicated by Mr Davie to his friend, who was still in London, and it appears to have lessened his anxiety about settling as a physician, and probably was one of the main causes of his laying aside all thoughts of that profession.   Perhaps, too, on a nearer view, he did not find that the practice of medicine would afford him that leisure for pursuing chemical and other scientific objects, which he fancied it would do when he saw things at a greater distance.   Whatever was the cause, it is certain that soon after his return to Edinburgh in summer 1750, he abandoned entirely his views of the practice of medicine, and resolved to apply himself to agriculture.

The motives which determined him in the choice of the latter cannot now be traced with certainty. He inherited from his father a small property in Berwickshire, and this might suggest to him the business of husbandry.   But we ought rather, I think, to look for the motives that influenced him, in the simplicity of his character, and the moderation of his views, than in external circumstances. To one who, in the maturity of understanding, has leisure to look round on the various employments which exercise the skill and industry of man, if his mind is independent and unambitious, and if he has no sacrifice to make to vanity or avarice, the profession of a farmer may seem fairly entitled to a preference above all others.   This was exactly the case of Dr Hutton, and he appears to have been confirmed in his choice by the acquaintance which he made about that time with Sir John Hall of Dunglass, a gentleman of the same county, a man of ingenuity and taste for science, and also much conversant with the management of country affairs.

As he was never disposed to do any thing by halves, he determined to study rural economy in the school which was then reckoned the best, and in the manner which is undoubtedly the most effectual. He went into Norfolk, and fixed his residence for some time in that country, living in the house of a farmer, who served both for his landlord and his instructor. This he did in 1752; and many years afterwards I have often heard him mention, with great respect, the name of John Dybold, at whose house he had lived with much comfort, and whose practical lessons in husbandry he highly valued. He appears, indeed, to have enjoyed this situation very much : the simple and plain character of the society with which he mingled suited well with his own, and the peasants of Norfolk would find nothing in the stranger to set them at a distance from him, or to make them treat him with reserve. It was always true of Dr Hutton, that to an ordinary man he appeared to be an ordinary man, possessing a little more spirit and liveliness, perhaps, than it is usual to meet with. These circumstances made his residence in Norfolk great-

ly to his mind, and there was accordingly no period
of his life to which he more frequently alluded, in
conversation with his friends ; often describing,
with singular vivacity, the rural sports and little
adventures, which, in the intervals of labour, formed the amusement of their society.

While his head-quarters were thus established in
Norfolk, he made many journeys on foot into different parts of England ; and though the main object of these was to obtain information in agriculture, yet it was in the course of them that, to amuse
himself on the road, he first began to study mineralogy or geology.   In a letter to Sir John Hall,
he says, that he was become very fond of studying
the surface of the earth, and was looking with anxious curiosity into every pit, or ditch, or bed of a
river that fell in his way ; " and that, if he did not
always avoid the fate of Thales, his misfortune was
certainly not owing to the same cause."   This letter is from Yarmouth ; it has no date, but it is plain
from circumstances, that it must have been written
in 1753.

What he learned in Norfolk made him desirous
of visiting Flanders, the country in Europe where
good husbandry is of the oldest date.   He accordingly set out on a tour in that country, early in
spring 1754, and, travelling from Rotterdam
through Holland, Brabant, Flanders, and Picardy,
he returned to England about the middle of sum-

mer. He appears to have been highly delighted with the garden culture which he found to prevail in Holland and Flanders, but not so as to under-value what he had learnt in England. He says, in a letter to Sir John Hall, written soon after his arrival in London, " Had I doubted of it before I set out, I should have returned fully convinced that they are good husbandmen in Norfolk."

Though his principal object in this excursion was to acquire information in the practice of husbandry, he appears to have bestowed a good deal of attention on the mineralogy of the countries through which he passed, and has taken notice in his Theory of the Earth of several of the observations which he made at that time.

About the end of the summer he returned to Scotland, and hesitated awhile in the choice of a situation where he might best carry into effect his plans of agricultural improvement. At last he fixed on his own farm in Berwickshire, and accordingly set about bringing it into order with great vigour and effect. A ploughman whom he brought from Norfolk set the first example of good tillage which had been seen in that district; and Dr Hutton has the credit of being one of those who introduced the new husbandry into a country where it has since made more rapid advances than in any other part of Great Britain.

From this time, till about the year 1768, he re-

sided, for the most part, on his farm, visiting Edinburgh, however, occasionally. The tranquillity of
rural life affords few materials for biographical description ; and an excursion to the north of Scotland, which he made in 1764, is one of the few incidents which mark an interval of fourteen years,
passed mostly in the retirement of the country. He
made this tour in company with Commissioner, afterwards Sir George Clerk, a gentleman distinguished for his abilities and worth, with whom Dr Hutton had the happiness to live in habits of the most
intimate friendship. They set out by the way of
Crieff, Dalwhinnie, Fort-Augustus, and Inverness ;
from thence they proceeded through East-Ross into Caithness, and returned along the coast by Aberdeen to Edinburgh. In this journey Dr Hutton's
chief object was mineralogy, or rather geology,
which he was now studying with great attention.

For several years before this period, Dr Hutton
was concerned in the sal-ammoniac work, which had
been actually established on the foundation of the
experiments already mentioned, but remained in Mr
Davie's name, only, till 1765 ; at that time a co-
partnership was regularly entered into, and the
work carried on afterwards in the name of both.

He now found that his farm was brought into
the regular order which good husbandry requires,
and that, as the management of it became more easy,
it grew less interesting. An occasion offering of

letting it to advantage, he availed himself of it.
About the year 1768 he left Berwickshire entirely,
and became resident in Edinburgh, giving his un-
divided attention, from that time, to scientific pursuits.

Among other advantages which resulted to him
from this change of residence, we must reckon that
of being able to enjoy, with less interruption, the
society of his literary friends, among whom were
Dr Black, Mr Russell Professor of Natural Philosophy, Professor Adam Fergusson, Sir George
Clerk, already mentioned, his brother Mr Clerk of
Elden, Dr James Lind, now of Windsor, and several others. Employed in maturing his views,
and studying nature with unwearied application, he
now passed his time most usefully and agreeably to
himself, but in silence and obscurity with respect
to the world. He was, perhaps, in the most en-
viable situation in which a man of science can be
placed. He was in the midst of a literary society
of men of the first abilities, to all of whom he was
peculiarly acceptable, as bringing along with him a
vast fund of information and originality, combined
with that gaiety and animation which so rarely accompany the profounder attainments of science.
Free from the interruption of professional avoca-
tions, he enjoyed the entire command of his own
time, and had sufficient energy of mind to afford
himself continual occupation.

A good deal of his leisure was now employed in the prosecution of chemical experiments. In one of these experiments, which he has no where mentioned himself, but which I have heard of from Dr Black, he discovered that mineral alkali is contained in zeolite. On boiling the gelatinous substance obtained from combining that fossil with muriatic acid, he found that, after evaporation, sea salt was formed. Dr Black did not recollect exactly the date of this experiment, but from circumstances judged that it was earlier than 1772. It is, if I mistake not, the first instance of an alkali being discovered in a stony body. The experiments of M. Klaproth and Dr Kennedy have confirmed this conclusion, and led to others of the same kind.

In 1774 he made a tour through part of England and Wales, of which I find no memorandum whatever among his papers. I know, however, that at this time he visited the salt-mines in Cheshire, and made the curious observation of the concentric circles marked on the roof of these mines, to which he has referred in his Theory of the Earth, as affording a proof that the salt rock was not formed from mere aqueous deposition. His friend Mr Watt of Birmingham accompanied him in his visit to the mines.

It was after returning to Birmingham from Cheshire, that he set out on the tour into Wales. One of the objects of this tour, as I learnt from himself, was to discover the origin of the hard gravel of granulated quartz, which is found in such vast abundance in the soil about Birmingham, and indeed over a great tract of the central part of England. This gravel is so unlike that which belongs to a country of secondary formation, that it very much excited his curiosity; and his present journey was undertaken with a view to find out whether, among the primitive mountains of Wales, there were any that might be supposed to have furnished the materials of it. In Wales, however, he saw none that could, with any probability, be supposed to have done so ; and he was equally unsuccessful in all the other parts he visited, till returning, at a small distance from Birmingham, the place from whence he had set out, he found a rock of the very kind which he had been in search of. It belongs to a body of strata apparently primary, which break out between Broomsgrove and Birmingham, and have all the characters of the indurated gravel in question. If, however, they have furnished the materials of that gravel, it seems probable that it has been through the medium of the red sandstone, which abounds in those countries. *

In 1777 Dr Hutton's first publication was given to the world, viz. a small pamphlet, entitled, *Con-*

---

* Illustrations of the Huttonian Theory, p. 375.

siderations on the Nature, Quality, and Distinctions of Coal and Culm.   This little work, an octavo pamphlet of thirty-seven pages, was occasioned by a question that had arisen, Whether the small coal of Scotland is the same with the culm of England? and, Whether, of course, like the latter, it is entitled, when carried coastwise, to an exemption from the duty on coal? Some of the small coal from the Frith of Forth, which had been carried to the northern counties for the purpose of burning lime, had been considered by the revenue officers as liable to the same duty with other coal, while the proprietors contended that it ought only to pay the lighter duty levied on culm. This was warmly disputed; and, after occupying for some time the attention of the Board of Customs in Scotland, was at last brought before the Privy Council.

Dr Hutton's pamphlet was intended to supply the information necessary for forming a judgment on this question. It is very ingenious and satisfactory, though perhaps, considering the purpose for which it was written, it is on too scientific a plan, and conducted too strictly according to the rules of philosophical analysis. It proves that culm is the small, or refuse, of the infusible, or stone coal, such as that of Scotland for the most part is; that the small of the fusible coal, by caking or uniting together, becomes equally serviceable with the large

OF DR JAMES HUTTON.

coal ; whereas the small of the infusible, by running down like loose sand, cannot be made to burn
in the ordinary way, and is useful but for few purposes, so that it has been properly exempted from
the usual duty on coal. A criterion is also pointed
out for determining when small coal is to be regarded as culm, and when it may be considered as
coal ;—if, when a handful of it is thrown into a
red-hot shovel, the pieces burn without melting
down or running together, it decidedly belongs to
the former.

In the conclusion, an exemption from duty was
obtained for the small coal of Scotland, when carried coastwise, and this regulation was owing in a
great degree to the satisfactory information contained in Dr Hutton's pamphlet. It was a step, also,
toward the entire abolition of those injudicious duties which had been so long levied on coal, when
carried by sea beyond a certain distance from its
native place. This abolition happened several years
after the period we are speaking of, much to the
benefit of the country, and to the credit of the administration under which it took place.

As Dr Hutton always took a warm interest in
whatever concerned the advancement of the arts,
particularly in his native country, he entered with
great zeal into the project of an internal navigation
between the Friths of Forth and Clyde. The comparative merit of the different plans, according to

which that work was to be executed, gave rise to a
good deal of discussion, and even of controversy.
In these debates Dr Hutton took a share, and
wrote several pieces, in which the grave and the ludicrous were both occasionally employed.   None of
these pieces have been published ; but the plan that
was in the end adopted was that in favour of which
they were written.   It is unnecessary, however, to
enter into the merits of a question which has long
ceased to interest the public.

From the time of fixing his residence in Edinburgh, Dr Hutton had been a member of the Philosophical Society known to the world by the three
volumes of physical and literary essays so much and
so justly esteemed. *   In that society he read several papers ; but it was during the time that elapsed
between the publication of the last of the volumes
just mentioned, and the incorporation of the Philosophical into the Royal Society of Edinburgh; which
last was established by a royal charter in 1783.
None of these papers have been published, except
one in the second volume of the Transactions of
the Royal Society, " On certain Natural Appear-

* The Philosophical Society was instituted about the year
1739.   The first volume of Essays was published in 1754 ;
the second in 1756 ; the third in 1771.   From the year
1777 to 1782, the meetings of the Society were pretty regular, much owing to the zeal of Lord Kames.   Mr Maclaurin
may be regarded as the founder of this Society.

ances of the Ground on the Hill of Arthur's Seat."

The institution of the Royal Society of Edinburgh had the good effect of calling forth from Dr Hutton the first sketch of a theory of the earth, the formation of which had been the great object of his life. From the date formerly mentioned, when he was yet a very young man, and making excursions on foot through the different counties of England, till that which we are now arrived at, a period of about thirty years, he had never ceased to study the natural history of the globe, with a view of ascertaining the changes that have taken place on its surface, and of discovering the causes by which they have been produced.

He had become a skilful mineralogist, and had examined the great facts of geology with his own eyes, and with the most careful and scrupulous observation. In the course of these studies he had brought together a considerable collection of minerals peculiarly calculated to illustrate the changes which fossil bodies have undergone. He had also carefully perused almost every book of travels from which any thing was to be learned concerning the natural history of the earth; and, in consequence both of reading and observation, was eminently skilled in physical geography.

If to all this it be added, that Dr Hutton was a good chemist, and possessed abilities excellently

adapted to philosophical research, it will be acknowledged, that few men have entered with better pre-
paration on the arduous task of investigating the
true theory of the earth.   Several years before the
time I am now speaking of, he had completed the
great outline of his system, but had communicated
it to very few ; I believe to none but his friends
Dr Black and Mr Clerk of Elden.   Though for-
tified in his opinion by their agreement with him,
(and it was the agreement of men eminently qualified to judge,) yet he was in no haste to publish his
theory ; for he was one of those who are much
more delighted with the contemplation of truth,
than with the praise of having discovered it.   It
might, therefore, have been a long time before he
had given any thing on this subject to the public,
had not his zeal for supporting a recent institution
which he thought of importance to the progress of
science in his own country induced him to come
forward, and to communicate to the Royal Society
a concise account of his theory of the earth.

As I have treated of this theory in a separate
essay, particularly destined to the illustration of it,
I shall here content myself with a very general
outline.

I. The object of Dr Hutton was not, like that
of most other theorists, to explain the first origin
of things.   He was too well skilled in the rules of

sound philosophy for such an attempt ; and he accordingly confined his speculations to those changes which terrestrial bodies have undergone since
the establishment of the present order, in as far as
distinct marks of such changes are now to be discovered.

With this view, the first general fact which he
has remarked is, that by far the greater part of the
bodies which compose the exterior crust of our
globe, bear the marks of being formed out of the
materials of mineral or organized bodies, of more
ancient date.   The spoils or the wreck of an older
world are every where visible in the present, and,
though not found in every piece of rock, they are
diffused so generally as to leave no doubt that the
strata which now compose our continents are all
formed out of strata more ancient than themselves.

II. The present rocks, with the exceptions of
such as are not stratified, having all existed in the
form of loose materials collected at the bottom of
the sea, must have been consolidated and converted into stone by virtue of some very powerful and
general agent.   The consolidating cause which he
points out is subterraneous heat, and he has removed the objections to this hypothesis by the introduction of a principle new and peculiar to himself.
This principle is the compression which must have
prevailed in that region where the consolidation of

mineral substances was accomplished. Under the weight of a superincumbent ocean, heat, however intense, might be unable to volatilize any part of those substances which, at the surface, and under the lighter pressure of our atmosphere, it can entirely consume. The same pressure, by forcing those substances to remain united, which at the surface are easily separated, might occasion the fusion of some bodies which in our fires are only calcined. Hence the objections that are so strong and unanswerable, when opposed to the theory of volcanic fire, as usually laid down, have no force at all against Dr Hutton's theory; and hence we are to consider this theory as hardly less distinguished from the hypothesis of the Vulcanists, in the usual sense of that appellation, than it is from that of the Neptunists, or the disciples of Werner.

III. The third general fact on which this theory is founded, is, that the stratified rocks, instead of being either horizontal, or nearly so, as they no doubt were originally, are now found possessing all degrees of elevation, and some of them even perpendicular to the horizon; to which we must add, that those strata which were once at the bottom of the sea are now raised up, many of them, several thousand feet above its surface. From this, as well as from the inflexions, the breaking and separation

of the strata, it is inferred, that they have been raised up by the action of some expansive force placed
under them.   This force, which has burst in pieces
the solid pavement on which the ocean rests, and
has raised up rocks from the bottom of the sea, into mountains 15,000 feet above its surface, exceeds
any which we see actually exerted, but seems to
come nearer to the cause of the volcano or the
earthquake than to any other, of which the effects
are directly observed.   The immense disturbance,
therefore, of the strata, is in this theory ascribed to
heat acting with an expansive power, and elevating
those rocks which it had before consolidated.

IV. Among the marks of disturbance in which
the mineral kingdom abounds, those great breaches
among rocks, which are filled with materials different from the rock on either side, are among the
most conspicuous.   These are the veins, and comprehend, not only the metallic veins, but also those
of whinstone, of porphyry, and of granite, all of
them substances more or less crystallized, and none
of them containing the remains of organized bodies.
These are of posterior formation to the strata
which they intersect, and in general also they carry
with them the marks of the violence with which
they have come into their place, and of the disturbance which they have produced on the rocks

already formed. The materials of all these veins
Dr Hutton concludes to have been melted by subterraneous heat, and, while in fusion, injected
among the fissures and openings of rocks already
formed, but thus disturbed, and moved from their
original place.

This conclusion he extends to all the masses of
whinstone, porphyry, and granite, which are interposed among strata, or raised up in pyramids, as
they often appear to be, through the midst of them.
Thus, in the fusion and injection of the unstrati-
fied rocks, we have the third and last of the great
operations which subterraneous heat has performed
on mineral substances.

V. From this Dr Hutton proceeds to consider
the changes to which mineral bodies are subject
when raised into the atmosphere. Here he finds,
without any exception, that they are all going to
decay; that from the shore of the sea to the top
of the mountain, from the softest clay to the
hardest quartz, all are wasting and undergoing
a separation of their parts. The bodies thus resolved into their elements, whether chemical or mechanical, are carried down by the rivers to the sea,
and are there deposited. Nothing is exempted
from this general law : among the highest mountains and the hardest rocks, its effects are most

clearly discerned; and it is on the objects which appear the most durable and fixed, that the characters of revolution are most deeply imprinted.

On comparing the first and the last of the propositions just enumerated, it is impossible not to perceive that they are two steps of the same progression, and that mineral substances are alternately dissolved and renewed. These vicissitudes may have been often repeated; and there are not wanting remains among mineral bodies, that lead us back to continents from which the present are the third in succession. Here, then, we have a series of great natural revolutions in the condition of the earth's surface, of which, as the author of this theory has remarked, we neither see the beginning nor the end; and this circumstance accords well with what is known concerning other parts of the economy of the world. In the continuation of the different species of animals and vegetables that inhabit the earth, we discern neither a beginning nor an end; and in the planetary motions, where geometry has carried the eye so far both into the future and the past, we discover no mark either of the commencement or termination of the present order. It is unreasonable, indeed, to suppose that such marks should any where exist. The Author of nature has not given laws to the universe, which, like the institutions of men, carry in themselves the elements of their own destruction; he has not permitted in his works any symptom of infancy or of old age, or any sign by which we may estimate either their future or their past duration. He may put an end, as he no doubt gave a beginning, to the present system, at some determinate period of time; but we may rest assured, that this great catastrophe will not be brought about by the laws now existing, and that it is not indicated by any thing which we perceive.

It would be desirable to trace the progress of an author's mind in the formation of a system where so many new and enlarged views of nature occur, and where so much originality is displayed. On this subject, however, Dr Hutton's papers do not afford so much information as might be wished for, though something may be learnt from a few sketches of an Essay on the Natural History of the Earth, evi·dently written at a very early period, and intended, it would seem, for parts of an extensive work, of which, as often happens with the first attempts to generalize, the plan was never executed, and may never have been accurately digested.

From these sketches it appears that the first of the propositions just enumerated, viz. that a vast proportion of the present rocks is composed of materials afforded by the destruction of bodies, animal, vegetable, and mineral, of more ancient formation, was the first conclusion that he drew from his observations.

The second seems to have been, that all the present rocks are without exception going to decay, and their materials descending into the ocean. These two propositions, which are the extreme points, as it were, of his system, appear, as to the order in which they became known, to have preceded all the rest. They were neither of them, even at that time, entirely new propositions, though, in the conduct of the investigation, and in the use made of them, a great deal of originality was displayed. The comparison of them naturally suggested to a mind not fettered by prejudice, nor swayed by authority, that they are two steps of the same progression ; and that, as the present continents are composed from the waste of more ancient land, so, from the destruction of them, future continents may be destined to arise. Dr Hutton accordingly, in the notes to which I allude, insists much on the perfect agreement of the structure of the beds of grit or sandstone, with that of the banks of unconsolidated sand now formed on our shores, and shows that these bodies differ from one another in nothing but their compactness and induration.

In generalizing these appearances, he proceeded a step farther, considering this succession of continents as not confined to one or two examples, but as indefinitely extended, and the consequence of laws perpetually acting. Thus he arrived at the

new and sublime conclusion, which represents nature as having provided for a constant succession of
land on the surface of the earth, according to a plan
having no natural termination, but calculated to
endure as long as those beneficent purposes, for
which the whole is destined, shall continue to exist.

This conclusion, however, was but a suggestion,
till the mechanism was inquired into by which this
grand renovation may be brought about, or by
which loose materials can be converted into stone,
and elevated into land.   This led to an investigation of the mineralizing principle, or the cause of
the consolidation of mineral bodies.   And Dr Hutton appears accordingly, with great impartiality,
and with no physical hypothesis whatever in his
mind, to have begun with inquiring into the nature
of the fluidity which so many mineral substances
seem to have possessed previous to the acquisition
of their present form.   After a long and minute
examination, he came to the conclusion, that the
fluidity of these substances has been what he terms
simple, that is to say, not such as is produced
by combination with a solvent.   The two general
facts from which this conclusion follows are, first,
that no solvent is capable of holding in solution all
mineral substances, nor even all such varieties
of them as are often united in the same specimen ;
and, secondly, that in the bodies composed of fragments of other bodies, the consolidation is so com-

plete that no room is left for a solvent to have ever occupied. The substance, therefore, which was the cause of the fluidity of mineral bodies, and prepared them for consolidation, must have been one that could act on them all, which occupied no space within them, and could find its way through them, whatever was the degree of their compactness and induration. Heat is the only substance which has these properties ; and is the only one, therefore, which, without manifest contradiction, can be assigned as the cause of mineral consolidation.

Many difficulties, however, were still to be removed before this hypothesis was rendered completely satisfactory ; but in what order Dr Hutton proceeded to remove them, the notes above mentioned do not enable me to state. We may nevertheless conjecture, with considerable probability, what the step was which immediately followed.

It must have occurred to him, as an objection to the consolidation of minerals by subterraneous heat, that many substances are found in the bowels of the earth in a state altogether unlike that into which they are brought by the action of our fires at the surface. Coal, for instance, by exposure to fire, has its parts dissipated ; the ashes which remain behind are a substance quite different from the coal itself ; and hence it would seem that this fossil can never before have been subjected to the action of fire. But is it certain, (we may suppose Dr Hutton to have said to himself,) if the heat had been applied to the coal in the interior of the earth, at the bottom of the sea, for example, that the same dissipation of the parts would have taken place? Would not the greater compression that must prevail in that region have prevented the dissipation, at least till a more intense heat was applied? And if the dissipation was prevented, might not the mass, after cooling, be very different from any thing that can be obtained by burning at the surface of the earth? It is plain that there is no reason whatever for answering these questions in the negative. And, on the contrary, if the analogy of nature is consulted, if the fact of water requiring more heat to make it boil when it is more compressed, or the experiments with Papin's digester, be considered, it will appear that the answer must be in the affirmative. Nay, it could not but seem reasonable to proceed a step farther, and, as the mixture of substances is known in so many instances to promote their fusibility, to suppose that, when the volatile parts of bodies were restrained, the whole mass might be reduced into fusion by heat, though, when these same parts were driven off, the residuum might be altogether infusible. Thus coal, when the charcoal and bitumen are forced to remain in union, may very well be a fusible substance, though, when the latter is permitted to escape, the former becomes one of the most refractory of all bodies.

12

In this way, and probably from this very instance, the effects of compression may have suggested themselves to Dr Hutton. He would soon perceive that the same principle could be very generally applied, and that it afforded the solution of a difficulty concerning limestone similar to that which has been just stated with respect to coal. Limestone is not found in the bowels of the earth having the causticity which it acquires by the action of fire, and hence one might conclude that it had never been exposed to the action of that element. But the experiments of Dr Black, before his friend was engaged in this geological investigation, * had proved that the causticity of lime depends on the expulsion of the aëriform fluid, since distinguished by the name of carbonic gas, which composes no less than two-fifths of the whole. This great discovery, which has extended its influence so widely over the science of chemistry, also led to important consequences in geology; and Dr Hutton inferred from it, that strong compression might prevent the caus-

---

* Dr Black's paper on magnesia, which contained this discovery, was communicated to the Philosophical Society of Edinburgh in June 1755, and was published in the second volume of their Essays, in the year following. Dr Hutton had at this time only begun his geological researches. It was not, I imagine, till after the year 1760 that they came to take the form of a theory.

ticity of lime, by confining the carbonic gas, even when great heat was applied, and that, as has been supposed of coal, the whole may have been melted in the interior of the earth, so as on cooling to acquire that crystallized or sparry structure which the carbonate of lime so frequently possesses. *

---

* In the view here presented of the principle of compression, as employed in the Huttonian Theory, it is considered as a hypothesis, conformable to analogy, assumed for the purpose of explaining certain phenomena in the natural history of the earth. It rests, therefore, as to its evidence, partly on its conformity to analogy, and partly on the explanation which it affords of the phenomena alluded to. In supposing that it derives probability from the last-mentioned source, we are far from assuming any thing unprecedented in sound philosophy. A principle is often admitted in physics, merely because it explains a great number of appearances ; and the theory of Gravitation itself rests on no other foundation.

The degree of this evidence will perhaps be differently appreciated, according to a man's habits of thinking, or the class of studies in which he has been chiefly engaged. To Dr Hutton himself it appeared very strong ; for he considered the fact of the liquefaction of mineral substances by heat as so completely established, that it affords a full proof of the fusibility of those substances having been increased by the compression which they endured in the bowels of the earth. In his view of the matter, no other proof seemed necessary, and he did not appear to think that the direct testimony of experiment, could it have been obtained, would have added much to the credibility of this part of his system.

For my part, I will acknowledge that the matter appears

It is unnecessary to carry our conjectures concerning the train of Dr Hutton's discoveries to a
greater length ; the developement of the principles
now enumerated, and the comparison of the results
with the facts observed in the natural history of
minerals, led to those discoveries, by a road that
will be easily traced by those who study his theory
with attention.

It might have been expected, when a work of so
much originality as this Theory of the Earth was
given to the world, a theory which professed to be
the result of such an ample and accurate induction,
and which opened up so many views, interesting
not to mineralogy alone, but to philosophy in general, that it would have produced a sudden and visi-

---

to me in a light somewhat different, and that, though the arguments just mentioned are sufficient to produce a very
strong conviction, it is a conviction that would be strengthened by an agreement with the results even of such experiments as it is within our reach to make. It seems to me,
that it is with this principle in geology, much as it is with
the parallax of the earth's orbit in astronomy ; the discovery
of which, though not necessary to prove the truth of the Copernican System, would be a most pleasing and beautiful addition to the evidence by which it is supported. So, in the
Huttonian geology, though the effects ascribed to compression are fairly deducible from the phenomena of the mineral kingdom itself, compared with certain analogies which
science has established, yet the testimony of direct experiment would make the evidence complete, and would leave
nothing that incredulity itself could possibly desiderate.

ble effect, and that men of science would have been
every where eager to decide concerning its real
value.   Yet the truth is, that it drew their attention
very slowly, so that several years elapsed before any
one showed himself publicly concerned about it,
either as an enemy or a friend.

Several causes probably contributed to produce
this indifference.   The world was tired out with
unsuccessful attempts to form geological theories,
by men often but ill informed of the phenomena
which they proposed to explain, and who proceeded also on the supposition that they could give an
account of the *origin* of things, or the first establishment of that system which is now the order of nature.   Men who guided their inquiries by a principle so inconsistent with the limits of the human
faculties, could never bring their speculations to a satisfactory conclusion, and the world readily enough
perceived their failure, without taking the trouble
to inquire into the cause of it.

Truth, however, forces me to add, that other
reasons certainly contributed not a little to prevent
Dr Hutton's theory from making a due impression
on the world.   It was proposed too briefly, and
with too little detail of facts, for a system which involved so much that was new, and opposite to the
opinions generally received.   The descriptions
which it contains of the phenomena of geology,
suppose in the reader too great a knowledge of the

things described. The reasoning is sometimes embarrassed by the care taken to render it strictly logical; and the transitions, from the author's peculiar notions of arrangement, are often unexpected and abrupt. These defects run more or less through all Dr Hutton's writings, and produce a degree of obscurity astonishing to those who knew him, and who heard him every day converse with no less clearness and precision, than animation and force. From whatever causes the want of perspicuity in his writings proceeded, perplexity of thought was not among the number; and the confusion of his ideas can neither be urged as an apology for himself, nor as a consolation to his readers.

Another paper from his pen, a Theory of Rain, appeared also in the first volume of the Edinburgh Transactions. He had long studied meteorology with great attention; and this communication contains one of the few speculations in that branch of knowledge entitled to the name of *theory*.

Dr Hutton begins with supposing that the quantity of humidity which air is capable of dissolving, increases with its temperature. Now, this increase must either be in the same ratio with the increase of heat, in a less ratio, or in a greater: in other words, for equal increments of heat, the increments of humidity must either constitute a series of which all the terms are equal to one another, or a series in which the terms continually decrease, or

one in which they continually increase. *   If either
of the two first laws was that which took place in
nature, a mixture of two portions of air, though
each contained as much humidity as it was capable
of dissolving, would never produce a condensation
of that humidity.   According to the first law, the
temperature, the humidity, and the power of containing humidity, in the mixture, being all arithmetical means between the same quantities, as they
existed previously to the mixture, the temperature
produced would be exactly that which was required
by the humidity to preserve it in its invisible form.
If the second law took place, the moisture actually
contained in the mixture would be less than the
temperature was capable of supporting ; so that, in-

---

* To speak strictly, the law which connects the increments
of humidity in the air with the increments of temperature,
is not confined to any one of the three suppositions here
made, but may involve them all.   The humidity dissolved
may be proportional to some *function* of the heat, that varies
in some places faster, and in others slower, than in the simple ratio of the heat itself.   Nevertheless, for that extent to
which observation reaches, the reasoning of Dr Hutton is
quite sufficient to prove that it varies faster ; or, in other
words, that if a curve be supposed, of which the abscissæ
represent the temperature, and the ordinates the humidity,
this curve, though it may in the course of its indefinite extent be in some places concave and in others convex toward
the axis, is wholly convex in all that part with which our
observations are concerned.

stead of a condensation of humidity, the air would become drier than before.

If, on the other hand, the third law be that which takes place, after the mixture of two portions of air of different temperatures, the humidity will be greater than the temperature is able to maintain, and therefore a condensation of it will follow. Now, the experience of every day proves, that the mixture of two portions of humid air of unequal temperatures, does indeed produce a condensation of moisture, and therefore we are authorized to conclude that the last-mentioned law is that which actually prevails. *

---

* It has been supposed that the chemical solution of humidity in air is necessarily implied in this theory of rain. The truth is, that the air is here considered only as the vehicle of the vapour, and that the transparent state of the latter is supposed to depend on the temperature, or the quantity of heat ; but whether that heat act on the vapour solely and directly, or indirectly, by increasing the power of the air to retain it in solution, is, with respect to this theory, altogether immaterial.

Dr Hutton has indeed used the common language concerning the solution of humidity in air ; but the supposition of such solution is not essential to his theory. He seemed, indeed, to entertain doubts about the reality of that operation, founded on the circumstance of evaporation taking place *in vacuo*. Experiments made by M. Dalton, since the death of Dr Hutton, show that there is great reason for supposing that the air has no chemical action whatever on the aqueous vapour contained in it.—Manchester Memoirs, Vol. V. p. 538.

It is obvious that this principle affords an explanation of the formation of clouds in the atmosphere,
and that currents of air, or winds, of different
temperatures, when they meet, must produce such
mixtures as have been described, and give rise consequently to the condensation of aqueous vapour.
When the supply of the humid air, entering into
the mixture, is continued, the quantity of cloud
formed will continually increase, and the small
globules of condensed moisture, uniting into drops,
must descend in rain.

But though we are thus in possession of a principle by which rain may be certainly produced, yet
whether it be the only one by which rain is produced may require some farther investigation.   Dr
Hutton accordingly, in order to determine this
point, has entered into a very ample detail concerning the rain under different climates, and in different regions of the earth.   The result is, that the
quantity of rain is, as nearly as can be estimated,
every where proportional to the humidity contained
in the air, and the causes which promote the mixture of different portions of air, in the upper
regions of the atmosphere.   Between the tropics,
for instance, the dry season is that in which the
uniform current of the trade-wind meets with
no obstruction in its circuit round the globe ; and
the rainy season happens when the sun approaches
to the zenith, and when the steadiness of the

trade-wind either yields to irregular variations,
or to the stated changes of the monsoons.

Thus, too, (to mention another extreme case,) in
certain countries distant from the sea, having little
inequality of surface, and exposed to great heat, no
rain whatever falls, and the sands of the desert are
condemned to perpetual sterility. Even there,
however, where a mountainous tract occurs, the
mixture of different portions of air produces a deposition of humidity ; perennial springs are found ;
and the fertile vales of Fezzan or Palmyra are exempted from the desolation of the surrounding
wilderness.

This ingenious theory attracted immediate attention, and was valued for affording a distinct notion of the manner in which cold acts in causing a
precipitation of humidity. It met, however, from
M. de Luc with a very vigorous and determined
opposition ; Dr Hutton defended it with some
warmth, and the controversy was carried on with
more sharpness, on both sides, than a theory in meteorology might have been expected to call forth.
For this Dr Hutton had least apology, if greatest
indulgence, on the score of temper, is due to the
combatant who has the worst argument. The
merits of the question cannot be considered here :
It is sufficient to remark, that they came ultimately to rest on a single point, Whether the refrigera-
tion of air is carried on by the mixture of the cold

and the hot air, or by the passage of the heat itself, without such mixture, from one portion of air to another. If the former holds, Dr Hutton's theory is established ; if the latter be true, M. de Luc's objections may at least merit examination.

Now, it is certain, that if not the only, yet almost the only, communication of heat through fluids, is produced by the mixture of one part of the fluid with another. The statical principle by which heat is thus propagated, was first, I believe, accurately explained by Dr Black, and since his time has been farther illustrated by the experiments of Count Rumford. These last have led their ingenious author to conclude that heat has no tendency to pass through fluids, otherwise than by the mixture of the parts of different temperature. The accuracy of this conclusion, in its full extent, may reasonably be questioned ; but this much of it is undoubtedly true, that when the particles of a body are at liberty to move freely among themselves, the direct communication of heat, compared with the statical, is evanescent, and may be regarded as a mere infinitesimal. M. de Luc's objections are therefore of no weight.

The Theory of Rain was republished by Dr Hutton in his Physical Dissertations several years afterwards, together with his answers to M. de Luc, and several other meteorological tracts, which contain many excellent examples of generalization, in a

branch of natural history where it is more easy to ac-
cumulate facts, and more difficult to ascertain principles, than in any other. *

---

* It may be proper to mention here some useful observations in meteorology which Dr Hutton made, but of which
he has given no account in any of his publications.

He was, I believe, the first who thought of ascertaining
the medium temperature of any climate by the temperature
of the springs. With this view he made a great number of
observations in different parts of Great Britain, and found,
by a singular enough coincidence between two arbitrary measures, quite independent of one another, that the temperature
of springs, along the east coast of this island, varies nearly at
the rate of a degree of Fahrenheit's thermometer for a degree
of latitude. This rate of change, though it cannot be general over the whole earth, is probably not far from the truth
for all the northern part of the temperate zone.

For estimating the effect which height above the level of
the sea has in diminishing the temperature, he also made a
series of observations at a very early period. By these observations he found that the difference between the state of
the thermometer in two places of a given difference of level,
and not very distant, in a horizontal direction, is a constant
quantity, or one which remains at all seasons nearly the
same, and is about $1°$ for 230 feet of perpendicular height.

I must, however, observe, that, on verifying these observations, I have found the rate of the decrease of temperature
a little slower than this, and very nearly a degree for 250
feet. This seems to hold for a considerable height above the
earth's surface, and will be found to come pretty near the
truth, to the height of five or six thousand feet. It is not,
however, probable, that the diminution of the temperature
is exactly proportional to the increase of elevation; and it

After the period of the two publications just
mentioned, Dr Hutton made several excursions

---

would seem, that at heights greater than the preceding, the
deviation becomes sensible ; the differences of heat varying in
a less ratio than the differences of elevation.

In explaining this diminution of temperature as we ascend
in the atmosphere, Dr Hutton was much more fortunate than
any other of the philosophers who have considered the same
subject. It is well known that the condensation of air converts part of the latent into sensible heat, and that the rarefac-
tion of air converts part of the sensible into latent heat. This
is evident from the experiment of the air-gun, and from many
others. If, therefore, we suppose a given quantity of air to be
suddenly transported from the surface to any height above it,
the air will expand on account of the diminution of pressure,
and a part of its heat becoming latent, it will become colder
than before. Thus also, when a quantity of heat ascends, by
any means whatever, from one stratum of air to a superior
stratum, a part of it becomes latent, so that an equilibrium
of heat can never be established among the strata ; but those
which are less must always remain colder than those that are
more compressed. This was Dr Hutton's explanation, and
it contains no hypothetical principle whatsoever.

To one who considers meteorology with attention, the
want of an accurate hygrometer can never fail to be a subject of regret. The way of supplying this deficiency which
Dr Hutton practised was by moistening the ball of a thermometer, and observing the degree of cold produced by the
evaporation of the moisture. The degree of cold, *cæteris
paribus,* will be proportional to the dryness of the air, and
affords, of course, a measure of that dryness. The same
contrivance, but without any communication whatsoever,
occurred afterwards to Mr Leslie, and being pursued

into different parts of Scotland, with a view of comparing certain results of his theory of the earth with
actual observation.　His account of granite, viz.
that it is a substance which, having been reduced
into fusion by subterraneous heat, has been forcibly
injected among the strata already consolidated, was
so different from that of other mineralogists, that it
seemed particularly to require farther examination.
He concluded, that if this account was just, some
confirmation of it must appear at those places where
the granite and the strata are in contact, or where
the former emerges from beneath the latter.　In
such situations, one might expect veins of the stone
which had been in fusion to penetrate into the stone
which had been solid ; and some imperfect descriptions of granitic veins gave reason to imagine that
this phenomenon was actually to be observed.　Dr
Hutton was anxious that an *instantia crucis* might
subject his theory to the severest test.

One of the places where he knew that a junction
of the kind he wished to examine must be found,
was the line where the great body of granite which
runs from Aberdeen westward, forming the central
chain of the Grampians, comes in contact with the
schistus which composes the inferior ridges of the

---

through a series of very accurate and curious experiments,
has produced an instrument which promises to answer all
the purposes of photometry as well as hygrometry, and so to
make a very important addition to our physical apparatus.

1

same mountains toward the south.   The nearest
and most accessible point of this line seemed likely
to be situated not far to the eastward of Blair in
Athol, and could hardly fail to be visible in the
beds of some of the most northern streams which
run into the Tay.   Dr Hutton having mentioned
these circumstances to the Duke of Athol, was in-
vited by that nobleman to accompany him in the
shooting season into Glentilt, which he did accordingly, together with his friend Mr Clerk of Elden,
in summer 1785.

The Tilt is, according to the seasons, a small
river, or an impetuous torrent, which runs through
a glen of the same name, nearly south-west, and
deeply intersects the southern ridges of the Grampian Mountains.   The rock through which its bed
is cut is in general a hard micaceous schistus ; and
the glen presents a scene of great boldness and asperity, often embellished, however, with the accom-
paniments of a softer landscape.

When they had reached the Forest Lodge, about
seven miles up the valley, Dr Hutton already found
himself in the midst of the objects which he wished
to examine.   In the bed of the river, many veins of
red granite (no less, indeed, than six large veins in
the course of a mile) were seen traversing the black
micaceous schistus, and producing, by the contrast
of colour, an effect that might be striking even to
an unskilful observer.   The sight of objects which

verified at once so many important conclusions in his system, filled him with delight ; and as his feelings, on such occasions, were always strongly expressed, the guides who accompanied him were convinced that it must be nothing less than the discovery of a vein of silver or gold, that could call forth such strong marks of joy and exultation.

Dr Hutton has described the appearances at this spot in the third volume of the Edinburgh Transactions, p. 79, and some excellent drawings of them were made by Mr Clerk, whose pencil is not less valuable in the sciences than in the arts. On the whole, it is certain, that of all the junctions of granite and schistus which are yet known, this at Glentilt speaks the most unambiguous language, and most clearly demonstrates the violence with which the granitic veins were injected among the schistus. *

---

* I must take this opportunity of correcting a mistake which I have made in describing the junction in Glentilt, (Illustrations of the Huttonian Theory, p. 310,) where I have said, that the great body of granite from which these veins proceed is not immediately visible. This, however, is not the fact, for the mountains on the north side of the glen are a mass of granite to which the veins can be directly traced. This I have been assured of by Mr Clerk. Dr Hutton has not described it distinctly ; and not having seen the union of the veins with the granite on the north side, when I visited the same spot, I concluded too hastily that it had not yet been discovered.

In the year following, Dr Hutton and Mr Clerk also visited Galloway, in search of granitic veins, which they found at two different places, where the granite and schistus come in contact. One of these junctions was afterwards very carefully examined by Sir James Hall and Mr Douglas, now Lord Selkirk, who made the entire circuit of a tract of granite country, which reaches from the banks of Loch Ken, where the junction is best seen, westward to the valley of Palnure, occupying a space of about 11 miles by 7. See Edinburgh Transactions, Vol. III. History, p. 8.

In summer 1787, Dr Hutton visited the island of Arran in the mouth of the Clyde, one of those spots in which nature has collected, within a very small compass, all the phenomena most interesting to a geologist. A range of granite mountains, placed in the northern part of the island, have their sides covered with primitive schistus of various kinds, to which, on the sea shore, succeed secondary strata of grit, limestone, and even coal. Here, therefore, Dr Hutton had another opportunity of examining the junction of the granite and schistus, and found abundance of the veins of the former penetrating into the latter. In three different places he met with this phenomenon; in the torrents that descend from the south side of Goatfield; in Glenrosa, on the west, and in the little river Sannax, on the north-east, of that mountain. From the

first of these he brought a specimen of some hun-
dred-weight, consisting of a block of schistus,
which includes a large vein of granite.

At the northern extremity of the island he had
likewise a view of the secondary strata lying upon
the primary, with their planes at right angles to
one another. In the great quantity, also, of pud-
ding-stone, containing rounded quartzy gravel, united by an arenaceous cement ; in the multitude of
whinstone dikes, which abound in this island ; and
in the veins of pitchstone, a fossil which he had not
before met with in its native place ; he found other
interesting subjects of observation ; so that he returned from this tour highly gratified, and used
often to say that he had no where found his expectations so much exceeded, as in the grand and instructive appearances with which nature has adorned this little island.

Mr John Clerk, the son of his friend Mr Clerk
of Elden, accompanied him in this excursion, and
made several drawings, which, together with a description of the island drawn up afterwards by Dr
Hutton, still remain in manuscript.

The least complete of the observations at Arran
was that of the junction of the primitive with the
secondary strata, which is but indistinctly seen in
that island, and only at one place. Indeed, the
contact of these two kinds of rock, though it forms
a line circumscribing the bases of all primitive

countries, is so covered by the soil, as to be visible in very few places. In the autumn of this same year, however, Dr Hutton had an opportunity of observing another instance of it in the bank of the river Jedd, about a mile above the town of Jedburgh. The schistus there is micaceous, in vertical plates, running from east to west, though somewhat undulated. Over these is extended a body of red sandstone, in beds nearly horizontal, having interposed between it and the vertical strata a breccia full of fragments of these last. Dr Hutton has given an account of this spot in the first volume of his Theory of the Earth, p. 432, accompanied with a copperplate, from a drawing by Mr Clerk.

In 1788 he made some other valuable observations of the same kind. The ridge of the Lammermuir Hills, in the south of Scotland, consists of primary micaceous schistus, and extends from St Abb's Head westward, till it join the metalliferous mountains about the sources of the Clyde. The sea coast affords a transverse section of this alpine tract at its eastern extremity, and exhibits the change from the primary to the secondary strata, both on the south and on the north. Dr Hutton wished particularly to examine the latter of these, and on this occasion Sir James Hall and I had the pleasure to accompany him. We sailed in a boat from Dunglass, on a day when the fineness of the weather permitted us to keep close to the foot of the rocks

which line the shore in that quarter, directing our
course southwards, in search of the termination of
the secondary strata. We made for a high rocky
point or head-land, the Siccar, near which, from
our observations on shore, we knew that the object
we were in search of was likely to be discovered.
On landing at this point, we found that we actually trode on the primeval rock, which forms alternately the base and the summit of the present
land. It is here a micaceous schistus, in beds nearly vertical, highly indurated, and stretching from
S. E. to N. W. The surface of this rock runs
with a moderate ascent from the level of low-water,
at which we landed, nearly to that of high-water,
where the schistus has a thin covering of red horizontal sandstone laid over it ; and this sandstone,
at the distance of a few yards farther back, rises into a very high perpendicular cliff. Here, therefore, the immediate contact of the two rocks is not
only visible, but is curiously dissected and laid open
by the action of the waves. The rugged tops of
the schistus are seen penetrating into the horizontal beds of sandstone, and the lowest of these last
form a breccia containing fragments of schistus,
some round and others angular, united by an arenaceous cement.

Dr Hutton was highly pleased with appearances
that set in so clear a light the different formations
of the parts which compose the exterior crust of the

8

earth, and where all the circumstances were combined that could render the observation satisfactory and precise. On us who saw these phenomena for the first time, the impression made will not easily be forgotten. The palpable evidence presented to us, of one of the most extraordinary and important facts in the natural history of the earth, gave a reality and substance to those theoretical speculations, which, however probable, had never till now been directly authenticated by the testimony of the senses. We often said to ourselves, What clearer evidence could we have had of the different formation of these rocks, and of the long interval which separated their formation, had we actually seen them emerging from the bosom of the deep? We felt ourselves necessarily carried back to the time when the schistus on which we stood was yet at the bottom of the sea, and when the sandstone before us was only beginning to be deposited, in the shape of sand or mud, from the waters of a superincumbent ocean. An epocha still more remote presented itself, when even the most ancient of these rocks, instead of standing upright in vertical beds, lay in horizontal planes at the bottom of the sea, and was not yet disturbed by that immeasurable force which has burst asunder the solid pavement of the globe. Revolutions still more remote appeared in the distance of this extraordinary perspective. *

* For a fuller deduction of the conclusions here referred

The mind seemed to grow giddy by looking so far
into the abyss of time ; and while we listened with
earnestness and admiration to the philosopher who
was now unfolding to us the order and series of
these wonderful events, we became sensible how
much farther reason may sometimes go than imagination can venture to follow. As for the rest, we
were truly fortunate in the course we had pursued
in this excursion ; a great number of other curious
and important facts presented themselves, and we
returned, having collected, in one day, more ample
materials for future speculation, than have sometimes resulted from years of diligent and laborious
research.

In the latter part of this same summer, (1788,)
Dr Hutton accompanied the Duke of Athol to the
Isle of Man, with a view of making a mineral survey of that island. What he saw there, however,
was not much calculated to illustrate any of the
great facts in geology. He found the main body
of the island to consist of primitive schistus, much
inclined, and more intersected with quartzy veins
than the corresponding schistus in the south of
Scotland. In two places on the opposite sides of
the island, this schistus was covered by secondary
strata, but the junction was no where visible. Some

---

to, see Theory of the Earth, Vol. I. p. 458 ; also Illustrations of the Huttonian Theory, p. 213.

granite veins were observed in the schistus, and many loose blocks of that stone were met with in the soil, or on the surface, but no mass of it was to be seen in its native place. The direction of the primitive strata corresponded very well with that in Galloway, running nearly from east to west. This is all the general information which was obtained from an excursion, which, in other respects, was very agreeable. Dr Hutton performed it in company with his friend Mr Clerk, and they again experienced the politeness and hospitality of the same nobleman who had formerly entertained them, on an expedition which deserves so well to be remembered in the annals of geology.

Though from the account now given, it appears that Dr Hutton's mind had been long turned with great earnestness to the study of the theory of the earth, he had by no means confined his attention to that subject, but had directed it to the formation of a general system, both of physics and metaphysics. *

---

* At what time these last speculations began to share his attention with the former, I have not been able to discover, though I have reason to believe, that before I became acquainted with him, which was about 1781, he had completed a manuscript treatise on each of them, the same nearly that he afterwards gave to the world. His speculations on general physics were of a date much earlier than this.

The Physical System referred to here forms the third part of a work entitled, Dissertations on different Subjects in Natural Philosophy, in one vol. 4to. 1792.

He tells us himself, that he was led to the study of
general physics, from those views of the properties
of body which had occurred to him in the prosecution of his chemical and mineralogical inquiries.
In those speculations, therefore, that extended so
far into the regions of abstract science, he began
from chemistry ; and it was from thence that he
took his departure in his circumnavigation both of
the material and intellectual world.

The chemist, indeed, is flattered more than any
one else with the hopes of discovering in what the
essence of matter consists ; and Nature, while she
keeps the astronomer and the mechanician at a great
distance, seems to admit him to more familiar converse, and to a more intimate acquaintance with her
secrets. The vast power which he has acquired
over matter, the astonishing transformations which
he effects, his success in analyzing almost all bodies,
and in reproducing so many, seem to promise that
he shall one day discover the essence of a substance
which he has so thoroughly subdued ; that he shall
be able to bind Proteus in his cave, and finally ex-
tort from him the secret of his birth ; in a word,
that he shall find out what matter is, of what elements it is composed, and what are the properties
essential to its existence

In entering upon this new inquiry, Dr Hutton
was forcibly struck with the very just reflection,
That we do by no means explain the nature of

body, when we describe it as made up of small particles; because, if we allow to these particles any
magnitude whatsoever, we do no more than affirm
that great bodies are made up of small ones.   The
elements of body must, therefore, be admitted to be
something unextended.   To these unextended elements Dr Hutton gave the name of Matter, and
carefully distinguished between that term and the
term Body, which he applied only to those combinations of matter that are necessarily conceived to
possess impenetrability, extension, and inertia.

The most accurate examination of the properties
of body confirms the truth of the opinion, that it is
composed of unextended elements.   Bodies may
be compressed into smaller dimensions; many by
the application of mechanical force, and all by the
diminution of their heat : nor is there any limit to
this compression, or any point beyond which the
farther reduction of volume becomes impossible.
This holds of substances the most compact, as well
as the most volatile and elastic, and clearly evinces
that the elements of body are not in contact with
one another, and that in reality we perceive nothing in body but the existence of certain powers
or forces, acting with various intensities, and in various directions.   Thus the supposed impenetrability, and of course the extension of body, is nothing
else than the effort of a resisting or repulsive
power; its cohesion, weight, &c. the efforts of

attractive power; and so with respect to all its other properties.

But if this be granted, and if it be true that in the material world every phenomenon can be explained by the existence of power, the supposition of extended particles as a *substratum* or residence for such power, is a mere hypothesis, without any countenance from the matter of fact. For if these solid particles are never in contact with one another, what part can they have in the production of natural appearances, or in what sense can they be called the residence of a force which never acts at the point where they are present? Such particles, therefore, ought to be entirely discarded from any theory that proposes to explain the phenomena of the material world.

Thus, it appears, that power is the essence of matter, and that none of our perceptions warrant us in considering even body as involving any thing more than force, subjected to various laws and modifications.

Matter, taken in this sense, is to be considered as indefinitely extended, and without inertia. Its presence through all space is proved by the universality of gravitation; and its want of inertia, by the want of resistance to the planetary motions. Thus, in our inquiry concerning physical causes, we are relieved from one great difficulty, that of supposing matter to act where it is not. The force of gravitation, according to this system, is not the action of two distant bodies upon one another, but it is the action of certain powers, diffused through all space, which may be transmitted to any distance. There seems to me, however, to remain a difficulty hardly less than that from which we appear to be relieved, viz. to assign a reason why the intensity with which such powers act on any body, should depend on the position and magnitude of all the bodies in the universe, and should bear to these continually the same relation. But, however this be, the ingenuity of Dr Hutton's reasonings cannot be questioned, nor, I think, the justness of many of his conclusions. His explanations of cohesion, heat, fluidity, deserve particular attention. In one thing, however, he seems to have fallen into an error, which runs through much of his reasoning, concerning the principles of gravitation and inertia. He affirms, that " without gravity, a body endowed with all the other material qualities would have no inertia ; that it would not diminish the velocity of the moving body by which it should be actuated, nor would it move a heavy body whatever were its velocity." * Now, this proposition, though from its nature it cannot be brought to the immediate test of experience, is certainly inconsistent with the principles of mechanics ; at the same time, it is true, that we would

---

* Dissertations, &c. p. 312, § 31.

not, in the case here supposed, have the same means of measuring the motion lost, or gained by collision, which we have in the actual state of bodies. This is perhaps what misled Dr Hutton ; and though his remarks on the measures of motion and force are very acute, and many of them very just, the mathematical reader will regret the want of that mode of reasoning, which has raised mechanics to so high a rank among the sciences.

It is impossible not to remark the affinity of this theory with that of the celebrated Boscovich, in which, as in this, all the phenomena of the material world are explained, by the supposition of forces variously modified, and without the assistance of solid or extended particles. These forces are supposed to be arranged round mathematical points, which are moveable, and act on one another by means of the forces surrounding them. A most ingenious application of this principle is made to all the usual researches of the mechanical philosophy, and, it must be confessed, that few theories have more beauty and simplicity to recommend them, or do better assist the imagination in the explanation of natural appearances. But it involves, in the whole of it, this great difficulty, that mathematical points are not only capable of motion, but capable of being endowed, or, at least, distinguished, by physical qualities. Dr Hutton, in his theory, has avoided this difficulty, by giving no other than a

negative definition of the Matter which he supposes
the elementary principle of body.   On this account,
though to the imagination his theory may want the
charms which the other possesses, yet it has the advantage of going just to the extent to which our
perceptions or our observations authorize us to proceed, and of being accurately circumscribed by the
limits pointed out by the laws of philosophical induction. *

---

* Though Boscovich's Theory was published long before
Dr Hutton's, so early, indeed, as the year 1758, there is no
reason to think that the latter was in any degree suggested
by the former.   Boscovich's theory was hardly known in
this country till about the year 1770, and the first sketches
of Dr Hutton's theory are of a much older date.   Besides,
the method of reasoning pursued by the authors is quite different ; and their conclusions, though alike in some things,
directly contrary in others, as in what regards gravity, inertia, &c.   The Monads of Leibnitz might more reasonably be
supposed to have pointed out to Dr Hutton the necessity of
supposing the elements of body to be unextended, if the originality of his own conceptions, and the little regard he paid
to authority in matters of theory, did not relieve us from the
necessity of looking to others for the sources of his opinions.
    The principal defect of his theory seems to me to consist
in this, that it does not state with precision the difference
between the constitution of those *powers* which simply form
matter, and those that form the more complex substance,
body.   In other words, it does not explain what must be
added to matter to make it body.   The answer seems to me
to be, that the addition of a repelling power, in all directions,

The existence of matter neither heavy nor inert, which he had taken so much pains to establish, was applied by him to explain the phenomena of light, heat, and electricity.   He considered all these three as modifications of the solar substance, and thought that many of the appearances they exhibit, are only to be explained on the supposition that they consist of an expansive force, of which inertness is not predicable ; in particular, that light is a power propagated from the sun in all directions, like gravity, with this difference, that it is repulsive, while gravity is attractive, and requires time for its transmission, which the latter does not, at least in any sensible quantity. *

The prosecution of this subject has led him to consider the nature of Phlogiston, a substance once so famous in chemistry, but of which the name has almost as entirely disappeared from the vocabulary of that science, as the word Vortex from the lan-

---

is sufficient for that purpose.   Such a repulsion, if strong enough, would produce both impenetrability and inertia. The matter, again, that possessed only an attractive power, like *gravity*, or a repulsive power only in a certain direction, like *light*, would not be inert nor impenetrable.   In this inference, however, from his system, I am not sure if I should meet with the author's approbation.

* See Dissertations V. and VI. on Matter and Motion, in the work above quoted. The chemical Dissertation on Phlogiston is in the same volume, p. 171.

guage of physical astronomy. The new and important experiments made on the calcination of metals, and on the composition of water, are, as is well known, the foundations of the antiphlogistic theory. Nobody was more pleased than Dr Hutton with these experiments, nor held in higher estimation the character and abilities of the chemists and philosophers by whom they were conducted. He was nevertheless of opinion, that the conclusions drawn from them are not altogether unexceptionable, nor deduced with a sufficient attention to every circumstance. This remark he thought peculiarly applicable to what regards the composition of water, to the phenomena of which experiment, the dissertation we are now speaking of is chiefly directed. The two aëriform fluids, it is there observed, which compose water, in order to unite, must not simply be brought together, for in that state they might remain for ever unchanged, but they must be set on fire, and made to burn, and from this burning there are evidently two substances which make their escape, namely, Light and Heat. Though, therefore, the weight of the water generated, and of the gases combined, may be admitted to be equal, yet it must be acknowledged that two substances are lost, which the chemist cannot confine in his closest vessel, nor weigh in his finest balance, and it is going much farther than we are authorized to do, either by experiment or

analogy, to conclude that these substances have had
no effect.   As heat and light, in Dr Hutton's system, are composed of that matter which does not
gravitate, the exact coincidence which M. Lavoisier observed between the weight of the water produced and that of the two elastic fluids, united in
the composition of it, was no argument, in his eyes,
against the escape of a very essential part of the in-
gredients.

Pursuing the same reasoning, he shows how little ground there is to suppose that the heat and
light evolved in this experiment proceed from the
vital air; and he concludes, that the real explanation of the process is, that by burning, the matter
of light and heat, or the phlogiston of the hydro-
genous gas, is set at liberty, and is thus enabled to
unite with the vital air.

In the same manner, on examining what relates
to the burning of inflammable bodies, he finds the
oxygenous gas unequal to the effect of furnishing
by its latent heat, or caloric, the whole of the sensible heat that is produced.   He concludes, therefore, that the hypothesis of the existence of phlogiston in those bodies that are termed inflammable, is necessary to account for the phenomena of
burning ; phenomena, as he justly remarks, which
are among the most curious and important of any
that are exhibited by the material world.   On the
whole, it cannot be doubted, that great ingenuity

and much sound argument are displayed through-
cut the whole of this dissertation, and that whatever be ultimately decided with regard to the principle for which the author so strenuously contends,
he has made it evident, that the conclusions of the
antiphlogistic theory have been drawn with too
much precipitancy, and carried farther than is war-
ranted by the strict rules of inductive philosophy.

The subject of Fire, Light, and Heat, was resumed by Dr Hutton several years after this period, and formed the subject of a series of papers
which he read in the Royal Society of Edinburgh,
and afterwards published separately. He there explains more fully his notion of the substances just
mentioned, which he considers as different modifications of the solar matter, alike destitute of inertness and of gravity.

A more voluminous work from Dr Hutton's pen
made its appearance soon after the Physical Dissertations, viz. An Investigation of the Principles
of Knowledge, and of the Progress of Reason from
Sense to Science and Philosophy, in three volumes
quarto.

He informs us himself of the train of thought by
which he was led to the metaphysical speculations
contained in these volumes. He had satisfied himself, by his physical investigations, that body is not
what it is conceived by us to be, a thing necessarily
possessing volume, figure, and impenetrability, but

merely an assemblage of powers, that by their action produce in us the ideas of these external qualities. His curiosity, therefore, was naturally excited to inquire farther into the manner in which
we form our conceptions of body, or into the nature
of the intercourse which the mind holds with those
things that exist without it. In pursuing this
inquiry, he soon became convinced, that magnitude,
figure, and impenetrability, are no otherwise perceived by the mind than colour, taste, and smell;
that is, that what are called the primary qualities of
body, are precisely on the same footing with the secondary, and are both conceptions of the mind,
which can have no resemblance to the external
cause by which those conceptions are produced.
The world, therefore, as conceived by us, is the
creation of the mind itself, but of the mind acted
on from without, and receiving information from
some external power. But though, according to
this reasoning, there be no resemblance between
the world without us, and the notions that we
form of it, though magnitude and figure, though
space, time, and motion, have no existence but
in the mind; yet our perceptions being consistent,
and regulated by constant and uniform laws, are as
much realities to us, as if they were the exact
copies of things really existing; they equally interest our happiness, and must equally determine our
conduct. They form a system, not dependent on

the mind alone, but dependent on the action which
certain external causes have upon it.   The whole
doctrine, therefore, of moral obligation, remains the
same in this system, and in that which maintains
the perfect resemblance of our ideas to the causes
by which they are produced.

Many philosophers have regarded our ideas
as very imperfect representations of external things ;
but Dr Hutton considers their perfect dissimilitude
as completely proved.   Plato has likened the mind
to an eye, so situated, as to see nothing but the
faint images of objects projected on the bottom of a
dark cave, while the objects themselves are entirely
concealed ; but he thinks, that by help of philosophy, the mental eye may be directed toward the
mouth of the cave, and may perceive the objects in
their true figure and dimensions.   But, with Dr
Hutton, the figures seen at the bottom of the cave
have no resemblance to the originals without ; nor
can man, by any contrivance, hold communication
with those originals, nor ever know any thing
about them, except that they are not what they
seem to be, and have no property in common with
the figures which denote their existence.   In a
word, external things are no more like the perceptions they give rise to, than wine is similar to intox-
ication, or opium to the delirium which it produces.

It has been already remarked, that this system,

however peculiar in other respects, involves in
it the same principles of morals with those more
generally received ; and the same may be said as to
the existence of God, and the immortality of the
soul. The view which it presents of the latter
doctrine deserves particularly to be remarked.
Death is not regarded here as the dissolution of a
connection between mind and that system of material organs, by means of which it communicated
with the external world, but merely as an effect of
the mind's ceasing to perceive a particular order or
class of things ; it is therefore only the termination
of a certain mode of thought ; and the extinction,
not of any mental power, but of a train of conceptions, which, in consequence of external impulse,
had existed in the mind. Thus, as nothing essential to intellectual power perishes, we are to consider death only as a passage from one condition of
thought to another ; and hence this system appeared, to the author of it, to afford a stronger argument than any other, for the existence of the
mind after death.

Indeed, Dr Hutton has taken great pains to deduce from his system, in a regular manner, the leading doctrines of morality and natural religion, having dedicated the third volume of his book almost
wholly to that object. It is worthy of remark, that
while he is thus employed, his style assumes a better tone, and a much greater degree of perspicuity,

than it usually possesses. Many instances might be pointed out, where the warmth of his benevolent and moral feelings bursts through the clouds that so often veil from us the clearest ideas of his understanding. One, in particular, deserves notice, in which he treats of the importance of the female character to society in a state of high civilization. * A felicity of expression, and a flow of natural eloquence, inspired by so interesting a subject, make us regret that his pen did not more frequently do justice to his thoughts.

The metaphysical theory, of which the outline (though very imperfectly) has now been traced, cannot fail to recal the opinions maintained by Dr Berkeley concerning the existence of matter. The two systems do indeed agree in one material point, but differ essentially in the rest. They agree in maintaining, that the conceptions of the mind are not copied from things of the same kind existing without it ; but they differ in this, that Dr Berkeley imagined that there is nothing at all external, and that it is by the direct agency of the Deity that sensation and perception are produced in the mind. Dr Hutton holds, on the other hand, that there is an external existence, from which the mind receives its information, and by the action of which impres-

---

* Investigation of the Principles of Knowledge, Vol. III. p. 588, &c.

sions are made on it ; but impressions that do not at all resemble the powers by which they are caused.

The reasonings, also, by which the two theories are supported, are very dissimilar, though perhaps they so far agree, that if Dr Berkeley had been better acquainted with physics, and had made it more a rule to exclude all hypothesis, he would have arrived precisely at the same conclusion with Dr Hutton. Indeed, I cannot help being of opinion, that every one will do so, who, in investigating the origin of our perceptions, determines to reason without assuming any hypothesis, and without taking for granted any of those maxims which the mind is disposed to receive, either, as some philosophers say, from habit, or, as others maintain, from an instinctive determination, (such as has been termed *common sense,)* that admits of no analysis. Though this may not be the kind of reasoning best suited to the subject, yet it is so analogous to what succeeds in other cases, that it is good to have an example of it, and, on that account, were it for nothing else, the theory we are now speaking of certainly merits more attention than it has yet met with. * The great size of the book, and the obscurity which may

* I have hardly found this work of Dr Hutton's quoted by any writer of eminence, except by Dr Parr, in his Spital Sermon, a tract no less remarkable for learning and acuteness, than for the liberality and candour of the sentiments which it contains.

justly be objected to many parts of it, have probably
prevented it from being received as it deserves, even
among those who are conversant with abstract speculation.   An abridgment of it, judiciously executed, so as to state the argument in a manner both
perspicuous and concise, would, I am persuaded,
make a valuable addition to metaphysical science.

The publication of this work was Dr Hutton's
occupation on his recovery from a severe illness,
with which he was seized in summer 1793.   Before
this time he had enjoyed a long continuance of
good health, and great activity both of body and
mind.   The disorder that now attacked him (a re-
tention of urine) was one of those that most immediately threaten life, and he was preserved only by
submitting to a dangerous and painful operation.
He was thus reduced to a state of great weakness,
and was confined to his room for many months.
By degrees, however, the goodness of his constitution, aided, no doubt, by the vigour and elasticity
of his mind, restored him to a considerable measure
of health, and rendered his recovery much more
complete than could have been expected.   One of
his amusements, when he had regained some tolerable degree of strength, was in superintending the
publication, and correcting the proof-sheets, of the
work just mentioned.

During his convalescence, his activity was farther
called into exertion, by an attack on his Theory of

the Earth, made by Mr Kirwan, in the Memoirs of the Irish Academy, * and rendered formidable, not by the strength of the arguments it employed, but by the name of the author, the heavy charges which it brought forward, and the gross misconceptions in which it abounded. †

Before this period, though Dr Hutton had been often urged by his friends to publish his entire work on the Theory of the Earth, he had con-

---

* This was not the first attack which had been made on his theory, for M. de Luc, in a series of letters, inserted in the Monthly Review for 1790 and 1791, had combated several of the leading opinions contained in it. To these Dr Hutton made no other reply, than is to be met with occasionally in the enlarged edition of his Theory, published four years afterwards. If I do not mistake, however, he intended a more particular answer, and actually sent one to the editors of the same Review, who refused to insert it. This, indeed, I do not state with perfect confidence, as I speak only from recollection, and would not, on that authority, bring a positive charge of partiality against men who exercise a profession in which impartiality is the first requisite. Supposing, however, the statement here given to be correct, an excuse is still left for the Reviewers ; they may say, that, in communicating original papers, as they do not act in their judicial capacity, they are not bound to dispense justice with their usual blindness and severity, but may be permitted to relax a little from the exercise of a virtue that is so often left to be its own reward.

† For a defence of Dr Hutton against the charges here alluded to, I must take the liberty of referring to the Illustrations of the Huttonian Theory, p. 119 and 125.

tinually put off the publication, and there seemed
to be some danger that it would not take place in
his own life time. The very day, however, after
Mr Kirwan's paper was put into his hands, he began the revisal of his manuscript, and resolved immediately to send it to the press. The reason he
gave was, that Mr Kirwan had in so many instances
completely mistaken, both the facts, and the reasonings in his Theory, that he saw the necessity of
laying before the world a more ample explanation
of them. The work was accordingly published, in
two volumes octavo, in 1795 ; and contained, besides what was formerly given in the Edinburgh
Transactions, the proofs and reasonings much more
in detail, and a much fuller application of the principles to the explanation of appearances. The two
volumes, however, then published, do not complete
the theory : a third, necessary for that purpose, remained behind, and is still in manuscript.

After the publication of the work just mentioned, he began to prepare another for the press, on a
subject which had early occupied his thoughts, and
had been at no time of his life entirely neglected.
This subject was husbandry, on which he had written a great deal, the fruit both of his reading and
experience ; and he now proposed to reduce the
whole into a systematic form, under the title of
Elements of Agriculture. This work, which he
nearly completed, remains in manuscript. It is

written with considerable perspicuity ; and though I can judge but very imperfectly of its merits, I can venture to say, that it contains a great deal of solid and practical knowledge, without any of the vague and unphilosophic theory so common in books on the same subject. In particular, I must observe, that where it treats of climate, and the influence of heat, in accelerating the maturity of plants, it furnishes several views that appear to be perfectly new, and that are certainly highly interesting.

The period, however, was now not far distant, which was to terminate the exertions of a mind of such singular activity, and of such ardour in the pursuit of knowledge. Not long after the time we are speaking of, Dr Hutton was again attacked by the same disorder from which he had already made so remarkable a recovery. He was again saved from the danger that immediately threatened him, but his constitution had materially suffered, and nothing could restore him to his former strength. He recovered, indeed, so far as to amuse himself with study, and with the conversation of his friends, and even to go on with the work on agriculture, which was nearly completed. He was, however, confined entirely to the house ; and in the course of the winter 1796-7, he became gradually weaker, was extremely emaciated, and suffered much pain, but still retained the full activity and acuteness of his mind. He constantly employed himself in reading and writing, and was particularly pleased with the third and fourth volumes of Saussure's *Voyages aux Alpes*, which reached him in the course of that winter, and became the last study of one eminent geologist, as they were the last work of another. On Saturday the 26th of March he suffered a good deal of pain ; but, nevertheless, employed himself in writing, and particularly in noting down his remarks on some attempts which were then making towards a new mineralogical nomenclature. In the evening he was seized with a shivering, and his uneasiness continuing to increase, he sent for his friend Mr Russell, who attended him as his surgeon. Before he could possibly arrive, all medical assistance was in vain : Dr Hutton had just strength left to stretch out his hand to him, and immediately expired.

Dr HUTTON possessed, in an eminent degree, the talents, the acquirements, and the temper, which entitle a man to the name of a philosopher. The direction of his studies, though in some respects irregular and uncommon, had been highly favourable to the developement of his natural powers, especially of that quick penetration, and that originality of thought, which strongly marked his intellectual character. From his first outset in science, he had pursued the track of experiment and observation, and it was not till after being long exercised in this school, that he entered on the field of general and abstract speculation. He combined accordingly, through his whole life, the powers of an accurate observer, and of a sagacious theorist, and was as cautious and patient in the former character, as he was bold and rapid in the latter.

Long and continued practice had increased his powers of observation to a high degree of perfection; so that, in discriminating mineral substances, and in seizing the affinities or differences among geological appearances, he had an acuteness hardly to be excelled. The eulogy so happily conveyed in the Italian phrase of *osservatore oculatissimo*, might most justly be applied to him; for, with an accurate eye for perceiving the characters of natural objects, he had in equal perfection the power of interpreting their signification, and of decyphering those ancient hieroglyphics which record the revolutions of the globe. There may have been other mineralogists, who could describe as well the fracture, the figure, the smell, or the colour of a specimen; but there have been few who equalled him in reading the characters, which tell not only what a fossil *is*, but what it *has been*, and declare the series of changes through which it has passed. His expertness in this art, the fineness of his observations, and the ingenuity of his reasonings, were truly admirable. It would, I am persuaded, be difficult to find in any of the sciences a better illustration of the profound maxims established by Bacon, in his Prerogativæ Instantiarum, than were often afforded by Dr Hutton's mineralogical disquisitions, when he exhibited his specimens, and discoursed on them with his friends. No one could better apply the luminous instances to elucidate the obscure, the decisive to interpret the doubtful, or the simple to unravel the complex. None was more skilful in marking the gradations of nature, as she passes from one extreme to another; more diligent in observing the continuity of her proceedings, or more sagacious in tracing her footsteps, even where they were most lightly impressed.

With him, therefore, mineralogy was not a mere study of names and external characters, (though he was singularly well versed in that study also,) but it was a sublime and important branch of physical science, which had for its object to unfold the connection between the past, the present, and the future conditions of the globe. Accordingly, his collection of fossils was formed for explaining the principles of geology, and for illustrating the changes which mineral substances have gone through, in the passage which, according to all theories, they have made, from a soft or fluid, to a hard and solid state, and from immersion under the ocean, to elevation above its surface. The series of these changes, and the relative antiquity of the different steps by which they have been effected, were the objects which he had in view to explain; and his cabinet, though well adapted to this end, with regard to other purposes was very imperfect. They who expect to find, in a collection, specimens of all the species, and all the varieties, into which a system of artificial arrangement may have divided the fossil kingdom, will perhaps turn fastidiously from one that is not remarkable either for the number or brilliancy of the objects contained in it. They, on the other hand, will think it highly interesting, who wish to reason concerning the natural history of minerals, and who are not less eager to become acquainted with the laws that govern, than with the individuals that compose, the fossil kingdom.

The loss sustained by the death of Dr Hutton was aggravated, to those who knew him, by the consideration of how much of his knowledge had perished with himself, and, notwithstanding all that he had written, how much of the light collected by a long life of experience and observation, was now completely extinguished. It is indeed melancholy to reflect, that with all who make proficiency in the sciences founded on nice and delicate observation, something of this sort must unavoidably happen. The experienced eye, the power of perceiving the minute differences, and fine analogies, which discriminate or unite the objects of science; and the readiness of comparing new phenomena with others already treasured up in the mind; these are accomplishments which no rules can teach, and no precepts can put us in possession of. This is a portion of knowledge which every man must acquire for himself, and which nobody can leave as an inheritance to his successor. It seems, indeed, as if nature had in this instance admitted an exception to the rule, by which she has ordained the perpetual accumulation of knowledge among civilized men, and had destined a considerable portion of science continually to grow up and perish with the individual.

A circumstance which greatly distinguished the intellectual character of the philosopher of whom we now speak, was an uncommon activity and ardour of mind, upheld by the greatest admiration of whatever in science was new, beautiful, or sublime. The acquisitions of fortune, and the enjoyments which most directly address the senses, do not call up more lively expressions of joy in other men, than hearing of a new invention, or being made acquainted with a new truth, would, at any time, do in Dr Hutton. This sensibility to intellectual pleasure was not confined to a few objects, nor to the sciences which he particularly cultivated: he would rejoice over Watt's improvements on the steam-engine, or Cook's discoveries in the South Sea, with all the warmth of a man who was to share in the honour or the profit about to accrue from them. The fire of his expression, on such occasions, and the animation of his countenance and manner, are not to be described; they were always seen with great delight by those who could enter into his sentiments, and often with great astonishment by those who could not.

With this exquisite relish for whatever is beautiful and sublime in science, we may easily conceive what pleasure he derived from his own geological speculations. The novelty and grandeur of the objects offered by them to the imagination, the simple and uniform order given to the whole natural history of the earth, and, above all, the views opened of the wisdom that governs nature, are things to which hardly any man could be insensible; but to him they were matter, not of transient delight, but of solid and permanent happiness. Few systems, indeed, were better calculated than his, to entertain their author with such noble and magnificent prospects; and no author was ever more disposed to consider the enjoyment of them, as the full and adequate reward of his labours.

The great range which he had taken in science has sufficiently appeared, from the account already given of his works. *   There were indeed hardly any sciences, except the mathematical, to which he had not turned his attention; and his neglect of these probably arose from this, that, at the time when his acquaintance with them should have commenced, his love of knowledge had already fixed itself on other objects. The aptitude of his mind for geometrical reasoning was, however, proved on many occasions. His theory of rain rests on mathematical principles, and the conclusions deduced from them are perfectly accurate, though by no means obvious. I may add, that he had an uncommon facility in comprehending the nature of mechanical contrivances; and, for one who was not a practical engineer, could form, beforehand, a very sound judgment concerning their effects.

Notwithstanding a taste for such various infor-

---

* He had studied with great care several subjects of which no mention is made above. One of these was the Formation, or, as we may rather call it, the Natural History of Language. A portion of his metaphysical work is dedicated to the Theory of Language, Vol. I. p. 574, &c.; and Vol. II. p. 624, &c. He read several very ingenious papers on the Written Language, in the Royal Society of Edinburgh. (See Transactions of the Royal Society of Edinburgh, Vol. II. Hist. p. 5, &c. The Chinese language, as an extreme case in the invention of writing, had greatly occupied his thoughts, and is the subject of several of his manuscripts.

mation, and a mind of such constant activity, he read but few speculative books, directing his attention chiefly to such as furnished the materials of speculation. Of voyages, travels, and books relating to the natural history of the earth, he had an extensive knowledge : he had studied them with that critical discussion which such books require above all others ; carefully collecting from them the facts that appeared accurate, and correcting the narratives that were imperfect, either by a comparison with one another, or by applying to them the standard of probability which his own observation and judgment had furnished him with. On the other hand, he bestowed but little attention on books of opinion and theory ; and while he trusted to the efforts of his mind for digesting the facts he had obtained from reading or experience, into a system of his own, he was not very anxious, at least till that was accomplished, to be informed of the views which other philosophers had taken of the same subject. He was but little disposed to concede any thing to mere authority ; and to his indifference about the opinions of former theorists, it is probable that his own speculations owed some part, both of their excellencies and their defects.

As he was indefatigable in study, and was in the habit of using his pen continually as an instrument of thought, he wrote a great deal, and has left behind him an incredible quantity of manuscript, though imperfect, and never intended for the press. Indeed, his manner of life, at least after he left off the occupations of husbandry, gave him such a command of his time, as is enjoyed by very few. Though he used to rise late, he began immediately to study, and generally continued busy till dinner. He dined early, almost always at home, and passed very little time at table ; for he ate sparingly, and drank no wine. After dinner he resumed his studies, or, if the weather was fine, walked for two or three hours, when he could not be said to give up study, though he might, perhaps, change the object of it. The evening he always spent in the society of his friends. No professional, and rarely any domestic arrangements, interrupted this uniform course of life, so that his time was wholly divided between the pursuits of science and the conversation of his friends, unless when he travelled from home on some excursion, from which he never failed to return furnished with new materials for geological investigation.

To his friends his conversation was inestimable ; as great talents, the most perfect candour, and the utmost simplicity of character and manners, all united to stamp a value upon it. He had, indeed, that genuine simplicity, originating in the absence of all selfishness and vanity, by which a man loses sight of himself altogether, and neither conceals what is, nor affects what is not. This simplicity

4

pervaded his whole conduct; while his manner, which was peculiar, but highly pleasing, displayed a degree of vivacity hardly ever to be found among men of profound and abstract speculation. His great liveliness, added to this aptness to lose sight of himself, would sometimes lead him into little eccentricities, that formed an amusing contrast with the graver habits of a philosophic life.

Though extreme simplicity of manner does not unfrequently impart a degree of feebleness to the expression of thought, the contrary was true of Dr Hutton. His conversation was extremely animated and forcible, and, whether serious or gay, full of ingenious and original observation. Great information, and an excellent memory, supplied an inexhaustible fund of illustration, always happily introduced, and in which, when the subject admitted of it, the witty and the ludicrous never failed to occupy a considerable place.—But it is impossible, by words, to convey any idea of the effect of his conversation, and of the impression made by so much philosophy, gaiety, and humour, accompanied by a manner at once so animated and so simple. Things are made known only by comparison, and that which is *unique* admits of no description.

The whole exterior of Dr Hutton was calculated to heighten the effect which his conversation produced. His figure was slender, but indicated activity; while a thin countenance, a high forehead, and a nose somewhat aquiline, bespoke extraordinary acuteness and vigour of mind. His eye was penetrating and keen, but full of gentleness and benignity ; and even his dress, plain, and all of one colour, was in perfect harmony with the rest of the picture, and seemed to give a fuller *relief* to its characteristic features. *

The friendship that subsisted between him and Dr Black has been already mentioned, and was indeed a distinguishing circumstance in the life and character of both. There was in these two excellent men that similarity of disposition which must be the foundation of all friendship, and, at the same time, that degree of diversity, which seems necessary to give to friends the highest relish for the society of one another.

They both cultivated nearly the same branches of physics, and entertained concerning them nearly the same opinions. They were both formed with a taste for what is beautiful and great in science ; with minds inventive, and fertile in new combinations. Both possessed manners of the most genuine simplicity, and in every action discovered the sincerity and candour of their dispositions ; yet they

---

* A portrait of Dr Hutton, by Raeburn, painted for the late John Davidson, Esq. of Stewartfield, one of his old and intimate friends, conveys a good idea of a physiognomy and character of face to which it was difficult to do complete justice.

were in many things extremely dissimilar. Ardour, and even enthusiasm, in the pursuit of science,
great rapidity of thought, and much animation,
distinguished Dr Hutton on all occasions. Great
caution in his reasonings, and a coolness of head
that even approached to indifference, were characteristic of Dr Black. On attending to their conversation, and the way in which they treated any
question of science or philosophy, one would say
that Dr Black dreaded nothing so much as error,
and that Dr Hutton dreaded nothing so much as
ignorance ; that the one was always afraid of going
beyond the truth, and the other of not reaching it.
The curiosity of the latter was by much the most
easily awakened, and its impulse most powerful and
imperious. With the former, it was a desire which
he could suspend and lay asleep for a time ; with
the other, it was an appetite that might be satisfied
for a moment, but was sure to be quickly renewed.
Even the simplicity of manner which was possessed
by both these philosophers, was by no means precisely the same. That of Dr Black was correct,
respecting at all times the prejudices and fashions
of the world ; that of Dr Hutton was more care-
less, and was often found in direct collision with
both.

From these diversities, their society was infinitely pleasing, both to themselves and those about
them. Each had something to give which the

other was in want of. Dr Black derived great
amusement from the vivacity of his friend, the
sallies of his wit, the glow and original turn of his
expression ; and that calmness and serenity of mind
which, even in a man of genius, may border on
languor or monotony, received a pleasing impulse
by sympathy with more powerful emotions.

On the other hand, the coolness of Dr Black,
the judiciousness and solidity of his reflections,
served to temper the zeal, and restrain the impetu-
osity of Dr Hutton. In every material point of
philosophy they perfectly agreed. The theory of
the earth had been a subject of discussion with them
for many years, and Dr Black subscribed entirely
to the system of his friend. In science, nothing
certainly is due to authority, except a careful examination of the opinions which it supports. It is
not meant to claim any more than this in favour of
the Huttonian Geology ; but they who reject that
system, without examination, would do well to consider, that it had the entire and unqualified approbation of one of the coolest and soundest reasoners
of which the present age furnishes any example.

Mr Clerk of Elden was another friend, with
whom, in the formation of his theory, Dr Hutton
maintained a constant communication. Mr Clerk,
perhaps from the extensive property which his family had in the coal-mines near Edinburgh, was
early interested in the pursuits of mineralogy. His

inquiries, however, were never confined to the objects which mere situation might point out, and, through his whole life, have been much more directed by the irresistible impulse of genius, than by the action of external circumstances. Though not bred to the sea, he is well known to have studied the principles of naval war with unexampled success; and though not exercising the profession of arms, he has viewed every country through which he has passed with the eye of a soldier as well as a geologist. The interest he took in studying the surface no less than the interior of the earth; his extensive information in most branches of natural history; a mind of great resource, and great readiness of invention; made him, to Dr Hutton, an invaluable friend and coadjutor. It cannot be doubted, that, in many parts, the system of the latter has had great obligations to the ingenuity of the former, though the unreserved intercourse of friendship, and the adjustments produced by mutual suggestion, might render those parts undistinguishable even by the authors themselves. Mr Clerk's pencil was ever at the command of his friend, and has certainly rendered him most essential service.

But it was not to philosophers and men of science only that Dr Hutton's conversation was agreeable. He was little known, indeed, in general company, and had no great relish for the enjoy-

ment which it affords ; yet he was fond of domestic
society, and took great delight in a few private circles, where several excellent and accomplished individuals of both sexes thought themselves happy to
be reckoned in the number of his friends.   In one
or other of these, he was accustomed almost every
evening to seek relaxation from the studies of the
day, and found always the most cordial welcome.
A brighter tint of gaiety and cheerfulness spread
itself over every countenance when the Doctor entered the room ; and the philosopher who had just
descended from the sublimest speculations of metaphysics, or risen from the deepest researches of geology, seated himself at the tea-table, as much disengaged from thought, as cheerful and gay, as the
youngest of the company.   These parties were de-
lightful, and, by all who have had the happiness to
be present at them, will never cease to be remembered with pleasure.

He used also regularly to unbend himself with a
few friends, in the little society alluded to in Professor Stewart's Life of Mr Smith, and usually
known by the name of the *Oyster Club*.   This
club met weekly ; the original members of it were
Mr Smith, Dr Black, and Dr Hutton, and round
them was soon formed a knot of those who knew
how to value the familiar and social converse of
these illustrious men.   As all the three possessed
great talents, enlarged views, and extensive infor-

mation, without any of the stateliness and formality
which men of letters think it sometimes necessary to
affect ; as they were all three easily amused,—were
equally prepared to speak and to listen,—and as the
sincerity of their friendship had never been darkened
by the least shade of envy ; it would be hard to find
an example, where every thing favourable to good
society was more perfectly united, and every thing
adverse more entirely excluded.   The conversation was always free, often scientific, but never di-
dactic or disputatious ; and as this club was much
the resort of the strangers who visited Edinburgh,
from any object connected with art or with science,
it derived from thence an extraordinary degree of
variety and interest.   It is matter of real regret
that it has been unable to survive its founders.

The simplicity of manner that has been already
remarked as so strikingly exemplified in Dr Hutton, was but a part of an extreme disinterestedness
which manifested itself in every thing he did.   He
was upright, candid, and sincere ; strongly attached to his friends ; ready to sacrifice any thing to
assist them ; humane and charitable.   He set no
great value on money, or, perhaps, to speak properly, he set on it no more than its true value ; yet,
owing to the moderation of his manner of life, and
the ability with which his friend Mr Davie conducted their joint concerns, he acquired considerable wealth.

He was never married, but lived with his sisters, three excellent women, who managed his domestic affairs ; and of whom, only one, Miss Isabella Hutton, remained to lament his death. By her his collection of fossils, about which he left no particular instructions, was presented to Dr Black ; who thought that he could not better consult the advantage of the public, or the credit of his friend, than by giving it to the Royal Society of Edinburgh, under the condition that it should be completely arranged, and kept for ever separate, for the purpose of illustrating the Huttonian Theory of the Earth.

FINIS.

BIOGRAPHICAL ACCOUNT

OF THE LATE

JOHN ROBISON, LL. D. F. R. S. Edin.

AND PROFESSOR OF NATURAL PHILOSOPHY IN THE

UNIVERSITY OF EDINBURGH.

BIOGRAPHICAL ACCOUNT

OF

JOHN ROBISON, LL. D. *

THE distinguished person who is the subject of
this memoir was born at Boghall, in the parish of
Baldernock, near Glasgow, in the year 1739.   His
father, John Robison, had been early engaged in
commerce in Glasgow, where, with a character of
great probity and worth, he had acquired considerable wealth, and, before the birth of his son, had
retired to the country, and lived at his estate of
Boghall.

His son was educated at the grammar school of
Glasgow.   We have no accounts of his earliest acquirements, but must suppose them to have been
sufficiently rapid, as he entered a student of Humanity, in the University of Glasgow, in November 1750, and in April 1756 took his degree in
Arts.

Several Professors of great celebrity adorned that

* From the Transactions of the Royal Society of Edinburgh, Vol. VII. (1815.)—ED.

University about this period.   Dr Simson was one
of the first geometers of the age ; and Mr Adam
Smith had just begun to explain in his lectures
those principles which have since been delivered
with such effect in the Theory of Moral Sentiments, and in the Wealth of Nations.   Dr Moore
was a great master of the Greek language, and
added to extensive learning a knowledge of the
ancient geometry, much beyond the acquirement of
an ordinary scholar.

Under such instructors, a young man of far inferior talents to those which Mr Robison possessed, could not fail to make great advancement.   He
used, nevertheless, to speak lightly of his early proficiency, and to accuse himself of want of application, but from what I have learnt, his abilities and
attainments were highly respected by his contemporaries, and he was remarked at a very early period for the ingenuity of his reasonings as well as
the boldness of his opinions.   According to his
own account, his taste for the accurate sciences was
not much excited by the pure Mathematics, and
he only began to attend to them, after he discovered their use in Natural Philosophy.

In the year following that in which he took his
degree, Dr Dick, who was joint Professor of Natural Philosophy with his father, died, and Mr
Robison offered himself to the old gentleman as a
temporary assistant.   He was recommended, as I

have been told, by Mr Smith, but was nevertheless judged too young by Mr Dick, as he was not yet nineteen. The object to which his father, a man of exemplary piety, wished to direct his future prospects, was the Church, to which, however, he was at this time greatly averse, from motives which do not appear; but certainly not from any dislike to the objects or duties of the Clerical Profession. It was very natural for him to wish for some active scene, where his turn for Physical, and particularly Mechanical Science, might be exercised, and the influence of those indefinite and untried objects, which act so powerfully on the imagination of youth, directed his attention toward London. Professor Dick and Dr Simson joined in recommending him to Dr Blair, Prebendary of Westminster, who was then in search of a person to go to sea with Edward, Duke of York, and to assist his Royal Highness in the study of Mathematics and Navigation. When Mr Robison reached London in 1758, he learnt that the proposed voyage was by no means fixed, and after passing some time in expectation and anxiety, he found that the arrangement was entirely abandoned. This first disappointment in a favourite object could not fail to be severely felt, and had almost made him resolve on returning to Scotland.

He had been introduced, however, to Admiral Knowles, whose son was to have accompanied the

Duke of York, and the Admiral was too conversant with Nautical Science, not to discover in him a genius strongly directed to the same objects. Though the scheme of the Prince's nautical education was abandoned, the Admiral's views with respect to his son remained unaltered, and he engaged Mr Robison to go to sea with him, and to take charge of his instruction. From this point it is, that we are to date his nautical as well as scientific attainments.

About the middle of February 1759, a fleet sailed from Spithead under the command of Admiral Saunders, intended to co-operate with a military force which was to be employed, during the ensuing summer, in the reduction of Quebec. Young Knowles, whom Mr Robison had agreed to accompany, was a midshipman on board the Admiral's ship, the Neptune of 90 guns; but in the course of the voyage, being promoted to the rank of Lieutenant in the Royal William of 80 guns, Mr Robison went with him on board that ship, and was there rated as a midshipman.

The fleet arrived on the coast of America in April; but it was not till the beginning of May that the entire dissolution of the ice permitted it to ascend the River St Lawrence, and that the active scene of naval and military operations commenced, which terminated so much to the credit of the British arms. A person whose seafaring life was to be limited to two years, may well be considered as fortunate, in witnessing, during that short period, a series of events so remarkable as those which preceded and followed the taking of Quebec. Though great armies were not engaged, much valour and conduct were displayed ; the leaders on both sides were men of spirit and talents ; and, on the part of the English, the most cordial co-operation of the sea and land forces was worthy of men animated by the spirit of patriotism, or the love of glory ; the fate also of the gallant leader, who fell in the moment of victory, and in the prime of life, by repressing the exultation of success, gave a deeper interest to the whole transaction.

Of the operations of this period Mr Robison was by no means a mere spectator. A hundred seamen, under the command of Lieutenant Knowles, were drafted from the Royal William into the Stirling Castle, the Admiral's ship. Mr Robison was of this party, and had an opportunity of seeing a great deal of active service. At this time, also, he was occasionally employed in making surveys of the river and the adjacent grounds ; a duty for which he was eminently qualified, both by his skill as a mathematician, and his execution as a draughtsman.

It is, however, much to be regretted, that his papers, whether memorandums or letters, give no account of the incidents of this period ; so that we are left to conclude, from the history of the times, what were the events in which he must have taken part, or to gather, from the imperfect recollection of his conversation, the scenes in which he was actually engaged.   I have heard him express great admiration at the cool intrepidity which he witnessed, when the fire-ships, sent down the stream against the English navy, at anchor in the river, seemed to present a wall of fire, extending from one bank to another, from which nothing that floated on the water could possibly escape.   Without the smallest alarm or confusion, the British sailors assailed this flaming battlement in their boats, grappled the ships which composed it, and towed them to the shore, where they burnt down quietly to the water's edge.

An anecdote which he also used to tell, deserves well to be remembered.   He happened to be on duty in the boat in which General Wolfe went to visit some of his posts, the night before the battle, which was expected to be decisive of the fate of the campaign.   The evening was fine, and the scene, considering the work they were engaged in, and the morning to which they were looking forward, sufficiently impressive.   As they rowed along, the General, with much feeling, repeated nearly the whole of Gray's Elegy (which had appeared not long before, and was yet but little known) to an officer who sat with him in the stern of the boat ; adding, as he concluded, that " he would prefer being the author of that poem to the glory of beating the French tomorrow."

Tomorrow came, and the life of this illustrious soldier was terminated, amid the tears of his friends, and the shouts of his victorious army. Quebec fell of course ; and soon afterwards the fleet under Admiral Saunders sailed for England. When they arrived on the coast, they were informed that the Brest fleet was at sea, and that Sir Edward Hawke was in search of it. Without waiting for orders, Admiral Saunders sailed to reinforce Hawke, but came too late, the celebrated victory over Conflans, in Quiberon Bay, having been obtained (on the 20th of November) a few days before he joined. Whether the Royal William accompanied the rest of the fleet on this occasion, I have not been able to learn. The body of General Wolfe was brought home in that ship, and was landed at Spithead on the 18th of November. From that date to the beginning of next year, I find nothing concerning the Royal William, when that ship, with the Namur and some others, under the command of Admiral Boscawen, sailed on an expedition to the Bay of Quiberon. On this service the Royal William remained between five and six months, having been twice sent to cruise off Cape Finisterre, for five weeks each time.

About this period, a series of letters from Mr

Robison to his father begins; and though the letters do not enter much into particulars, they leave
us less at a loss about the remaining part of his seafaring life.

On the 3d of August the Royal William returned to Plymouth, the greater part of the crew being
totally disabled by the sea-scurvy, from which Mr
Robison himself had suffered very severely.   He
writes to his father, that, out of seven hundred and
fifty able seamen, two hundred and eighty-six were
confined to their hammocks, in the most deplorable
state of sickness and debility, while one hundred
and forty of the rest were unable to do more than
walk on deck.   This circumstance strongly marks,
to us, who have lately witnessed the exertions of
British sailors, in the blockade of Brest, and other
ports of the enemy, the improvement made in the
art of preserving the health of seamen within the
last fifty years.   The Royal William, notwithstanding the state of extreme distress to which her crew
was reduced, by a continuance at sea, of hardly six
months, was under the command of Captain Hugh
Pigott, one of the most skilful officers of the British navy.   Mr Robison, indeed, never at any
time mentioned his name without praise, for his
knowledge of seamanship, and the address with
which he used to work the ship, in such bad weather,
as rendered her almost unmanageable to the other
officers.   The art of preserving the health of the

seaman is a branch of nautical science, which had at that time been little cultivated. Our great Circumnavigator had not yet shown, that a ship's crew may sail round the globe, with less mortality than was to be expected in the same number of men, living for an equal period in the most healthful village of their native country.

Mr Robison's letters to his father, about this time, are strongly expressive of his dislike to the sea, and of his resolution to return to Glasgow, and to resume his studies, particularly that of theology, with a view of entering into the church. These resolutions, however, were for the present suspended, by a very kind invitation from Admiral Knowles, to come and live with him in the country, and to assist him in his experiments; " Thus, (says the admiral,) we shall be useful to one another." What these experiments were is not mentioned, but they probably related to ship-building, a subject which the admiral had studied with great attention. Mr Robison, accordingly, continued to enjoy a situation, and an employment, that must both have been extremely agreeable to him, till the month of February in the year following, when Lieutenant Knowles was appointed to the command of the Peregrine sloop of war, of 20 guns. Whether the plan of nautical instruction which Mr Robison proposed for his pupil was not yet completed, or whether he had, after all, come to a resolution

of pursuing a seafaring life, (of which there is
an appearance in some of his letters,) he embarked
in the Peregrine, and he even mentions his hopes
of being made purser to that ship. The first
service in which Captain Knowles was employed,
was to convoy the fleet to Lisbon. In a letter
from Plymouth, where they were forced in by the
weather, Mr Robison paints, in strong colours, the
difference between sailing in a small ship, like the
Peregrine, and a first rate, like the Royal William,
and the uncomfortable situation of all on board,
during a gale which they had experienced in coming down the Channel. The voyage, however,
gave him an opportunity of visiting Lisbon, on
which the traces of the earthquake were yet deeply
imprinted ; and the ship continuing to cruize off
the coast of Spain and Portugal, he had occasion to
land at Oporto, and other places on the Portuguese
coast. In the month of June he returned to
England ; and from this time quitted the navy,
though he did not give up hopes of preferment.
He returned to live with Admiral Knowles, and
in the end of the same summer, was recommended
by him to Lord Anson, the First Lord of the Admiralty, as a proper person to take charge of Harrison's time-keeper, which, at the desire of the
Board of Longitude, was to be sent, on a trial voyage, to the West Indies.

The ingenious artist just named had begun the

construction of his chronometer, on new principles, as early as the year 1726, and with the fortitude and patience characteristic of genius, had for thirty-five years struggled against the physical difficulties of his undertaking, and the still more discouraging obstacles which the prejudice, the envy, or the indifference of his contemporaries, seldom fail to plant in the way of an inventor. Notwithstanding all these, he had advanced constantly from one degree of perfection to another, and it was his fourth time-keeper, reduced to a portable size, and improved in all other respects, that was now submitted to examination. It was intended that Mr Robison should accompany young Harrison and the time-keeper, in a frigate, the Deptford, to Port Royal in Jamaica, in order to determine, on their landing, the difference of time, as given by the watch, and as found by astronomical observation. The time-keeper, accordingly, was put into the hands of Mr Robertson of the Naval School at Portsmouth, who determined its rate, from nine days that it remained in his custody, to be $2\frac{2}{3}''$ slow per day, and also, the error to be $3''$ slow on the 6th of November at noon, according to mean solar time.

The Deptford sailed on the 18th of November, and arrived at Port Royal on the 19th of January; on the 26th, Mr Robison observed the time of noon, and found it to answer to $4^{h.}$ $59'$ $7\frac{1}{2}''$ by the watch, and this being corrected for the error of

three seconds, and also for the daily accumulation
of $2\frac{2}{3}''$ for eighty-one days five hours, the interval
between the observations, the difference of longitude between Portsmouth and Port Royal came out
$5^{\text{h.}}$ $2'$ $47''$, only four seconds less than it was known
to be from other observations.

The instructions of the Board farther required,
that, as soon as an opportunity could be found, the
same two gentlemen should return with the watch
to Portsmouth, that, by a comparison of it with the
time there, the total error, during both voyages,
might be ascertained. The opportunity of return
occurred sooner than they had any reason to expect; for the Spanish war having now broke out,
an alarm of an invasion of Jamaica from St Do-
mingo, occasioned the governor to dispatch the
Merlin sloop of war to England, to give intelligence
of the danger. Mr Robison and Mr Harrison obtained leave to return in the Merlin, and sailed on
the 28th, having been but a few days in Jamaica.
This voyage was an epitome of all the disasters,
short of shipwreck, to which seafaring men are exposed. They experienced a continuation of the
most tempestuous weather, and the most contrary
winds, from the moment they quitted the Bahamas,
till they arrived at Spithead. To add to their distress, the ship sprung a great leak, three hundred
leagues from any land, and it required the utmost
skill and exertion to keep her from sinking.   In a

terrible gale, on the 14th of March, their rudder broke in two, so that they could no longer keep the ship's head to the wind ; and if the gale had not speedily moderated, they must inevitably have perished. When the voyage was near a conclusion, and they were congratulating themselves on the end of their troubles, the ship was found to be on fire, and the flames were extinguished with great difficulty. They reached Portsmouth on the 26th of March, and on the 2d of April, the mean noon by the watch was found to be at $11^{\text{h.}}$ $51'$ $31\frac{1}{2}''$, and, making correction for the error and rate, this amounted to $11^{\text{h.}}$ $58'$ $6\frac{1}{2}''$, so that the whole error, from the first setting sail, was only $1'$ $53\frac{1}{2}''$, which, in the latitude of Portsmouth, would not amount to an error, in distance, of twenty miles.

When Mr Robison undertook the voyage to Jamaica, he made no stipulation for any remuncration ; and Lord Anson assured him, that he should have no reason to repent the confidence which he placed in the Board. But when, on his return, he came to look for the reward, to which the success and trouble of the undertaking certainly entitled him, he soon found that he had greatly erred in leaving himself so much at the mercy of unforeseen contingencies. Lord Anson was ill of the disease of which he died, and was not in a condition to attend to business. Admiral Knowles was disgusted with the Admiralty, and with the Ministry, by

which he thought himself ill-used ; so that Mr Robison had nothing to look for from personal kindness, and could trust only to the justice and moderation of his claims. These were of little advantage
to him ; for such was the inattention of the Lords
of the Admiralty, and the members of the Board
of Longitude, that he could not obtain access to
any of them, nor even receive from them any answer to his memorials.

The picture which his letters to his father present, at this time, is that of a mind suffering severely
from unworthy treatment, where it was least suspected. Men in office do not reflect, while they
are busy about the concerns of nations, how much
evil may be done by their neglect to do justice to an
individual. They may be extinguishing the fire of
genius, thrusting down merit below the level it
should rise to, or prematurely surrounding the
mind of a young man with a fence of suspicion and
distrust, worse than the evils which it proposes to
avert. Like other kinds of injustice, this may,
however, meet with its punishment ; though the
victim of unmerited neglect may remain for ever
obscure, and his sufferings for ever unknown, he
may also emerge from obscurity, and the treatment
he has met with may meet the eye of the public.
It is probable that the member of these Boards
most conspicuous for rank or for science, would
not have been above some feeling of regret, if he

had learnt, that the young man, whose petitions he disregarded, was to become the ornament of his country, and the ill treatment he then met with, a material fact in the history of his life.

But though we must condemn the neglect of which Mr Robison had so much reason to complain, we do by no means regret that the recompense, which he or his friends had in view, was not actually conferred on him. This was no other than an appointment to the place of a purser in a ship of war; a sort of preferment which, to a man of the genius, information, and accomplishment of Mr Robison, must have turned out rather as a punishment than a reward. It was, however, the object which, by the advice of Sir Charles Knowles, he now aspired to; and, indeed, he had done so ever after his first voyage in the Royal William; for it appears that he had wished to be made purser to the Peregrine at the time when Lieutenant Knowles was appointed to the command of that ship, though, considering its smallness, the situation could have been attended with little emolument. *

---

* It is, however, true, that the place of Purser was afterwards offered to Mr Robison, but such a one as he could have no temptation to accept. In 1763, when Lord Sandwich was First Lord of the Admiralty, his solicitations were so far listened to, that he was appointed to the Aurora, of 40 guns, then on the stocks. As the ship must be long of being in

Thus disappointed in his hopes, Mr Robison resolved on returning to Glasgow, in order to quali-
fy himself for entering into the Church.   Indeed,
the idea of prosecuting his original destination seems
often to have occurred to him, even when his views
appeared to have a very different direction.   When
he left the Royal William in 1761, he was not without serious intentions of resuming the study of
theology.   This appears, both from a letter he
wrote to his father, about that time, and from one
which he himself received from young Knowles,
who rallies him on his new profession, and on the
singularity of having acquired a taste for theological
studies in the ward room of a man of war.   When
he undertook the voyage to Jamaica, he would have
wished to have had the patronage of his employers,
for obtaining some ecclesiastical preferment rather
than naval ; and only agreed to the latter, as it lay
more in the way of the Board of Longitude to help
one to promotion in the navy than in the church.   It
appears, that he had never ceased te express to Dr
Blair a desire of assuming the clerical character ;
and he actually had, from that gentleman, the offer
of a curacy in a living of bis own, to which, however,
the emolument annexed was so small, that, after

commission, and the pay of the purser, in the mean time,
very inconsiderable, Mr Robison declined accepting this appointment.

consultation with his father, he declined accepting
of it.

But, however Mr Robison's views may have varied, to one object he steadily adhered, viz. the cultivation of science, and the acquisition of whatever
knowledge the situations he was placed in brought
within his reach.

He returned, therefore, to Glasgow ; and a man,
whose object was the prosecution of science, could
not arrive at any place in a more auspicious moment,
as that city was about to give birth to two of the
greatest improvements, which, in the eighteenth
century, have distinguished the progress either of
the sciences or the arts. The one of these was the
discovery of Latent Heat, by the late Dr Black ;
the other was the invention of what may be properly called a New Steam-engine, by Mr Watt. The
former of these eminent men was then the Lecturer
on Chemistry in the University, and had just been
led, by a train of most ingeniously contrived experiments, to the knowledge of a principle which
seemed to promise better for an explanation of the
process which takes place when heat is communicated to bodies, than any thing yet known in chemistry, viz. that when water passes from a solid to a
fluid state, as much of its heat disappears as would
have raised its temperature, had it remained solid,
140 degrees higher than that which it actually possesses. Mr Robison was already known to Dr

Black, having been introduced to him before he left Glasgow ; but at that time he had not studied chemistry, to which, however, he was now bending his attention. He had the advantage of being initiated in it by the author of the discovery just mentioned; and the new views struck out by his master, did not fail to interest him in a study, which, from that time, came to occupy a new place in physical science.

Mechanics had always been his favourite pursuit, and his turn to whatsoever was connected with it, had brought him to be acquainted with Mr Watt before 1758, when he left the University. Mr Watt, who, at that time, exercised the profession of a mathematical instrument maker, was employed in fitting up the astronomical instruments bequeathed to the Observatory by the late Dr Macfarlane of Jamaica. Mr Robison, on his return, found him still residing in Glasgow, and exercising the same profession, and their former intimacy was naturally renewed. In 1764, an occurrence such as to an ordinary man would have been of no value, gave rise to the improvement of the steam-engine. A model of the common engine, Newcomen's, which belonged to the Natural Philosophy Class, was put into Mr Watt's hands in order to be repaired. As the model worked faster than the large engines, it was found impossible to supply it with steam, and it was in the attempt to obviate this difficulty, and in trying to produce a more perfect vacuum, that the idea of condensing the steam in a separate vessel first occurred to him. At the same time, by a curious coincidence, his experiments led him to conclusions concerning the great quantity of heat contained in steam, that were only to be explained on the principle of latent heat. Mr Robison lived in a state of great intimacy with Mr Watt, and was so much acquainted with the first steps of this invention, that his evidence on the subject of the originality of it was afterwards of great use in ascertaining the justness of his claim.

There could not be a better school for philosophical invention than Mr Robison enjoyed at this time, and, accordingly, he used always to say, that it was not till his second residence at Glasgow that he applied to study with his whole mind.

Dr Black was elected Professor of Chemistry in the University of Edinburgh in the summer of 1766; and, on leaving Glasgow, recommended Mr Robison as his successor. He was accordingly made choice of, and began his first course of chemical lectures in October 1766. He was appointed for one year only, but his success assured his continuance without any other limit than such as depended on himself.

He had also the charge of the education of the late Mr Macdowal of Garthland, and of Mr

Charles Knowles, a son of the Admiral. But of the particulars, during four years, about this time, I have been able to obtain little information.

The friendship of Admiral Knowles had been all along exerted toward Mr Robison, with an extraordinary degree of zeal and assiduity, and was now the means of procuring for him a very unlooked-for preferment, which removed him from his academical duties at Glasgow. The Empress of Russia, convinced of the importance of placing her marine on the best footing, made an application to the Government of this country, for permission to engage in her service some of the most able and experienced of our naval officers, to whom she might entrust both the contrivance and the execution of the intended reformation. The request was agreed to, and the person recommended was Admiral Sir Charles Knowles, who had long applied, with great diligence, to the study of naval architecture, as well as to that of every branch of his profession ; and who, about fifty years before, had been sent to Portugal on a similar mission. A proceeding so free from that jealousy which often marks the conduct of great nations no less than the dealings of the most obscure corporations, is particularly deserving of praise. From the first moment that this offer was made to the Admiral, he communicated it to Mr Robison, whom he wished to engage as his Secretary, and to whom, as he says in his letters, he looked for much assistance in the duty he was about to undertake. A very handsome appointment was made for Mr Robison, and in the end of December 1770, he set out with Sir Charles and his family on the journey to St Petersburgh, over land.

Admiral Knowles held the office of President of the Board of Admiralty; and his intention was, that Mr Robison should have the place of Secretary. The Russian Board, however, being constituted more on the plan of the French than the English, there was no place corresponding to that of our Secretary of the Admiralty. Mr Robison continued, therefore, in the character of Private Secretary to the Admiral.

During the first year of the Admiral's residence in Russia, and for the greater part of the second, Mr Robison remained with him, employed in forming and digesting a plan for improving the methods of building, rigging, and navigating the Russian ships of war, and for reforming, of consequence, the whole detail of the operations in the naval arsenals of that empire.

These innovations, however, met with more resistance than either the admiral or his secretary had permitted themselves to suppose. The work of reform, conducted by a foreigner, even when he is supported by despotic power, must proceed but slowly; jealousy, pride, and self-interest, will continually counteract the plans of improvement, and, by their vigilance and unceasing activity, will never wholly fail of success. All this was experienced by Admiral Knowles; yet there is no doubt that material advantages were derived, by the Russian navy, from the new system which he was enabled, partially, to introduce.

Mr Robison, from his first arrival at St Petersburgh, had applied with great diligence to the study of the Russian language, and had made himself so much master of it, as to speak and write it with considerable facility. In summer 1772, a vacancy happening in the mathematical chair attached to the Imperial Sea Cadet Corps of Nobles, at Cronstadt, Mr Robison was solicited to accept of that office. His nautical and mathematical knowledge qualified him singularly for the duties of it, and his proficiency in the Russian language removed the only objection that could possibly be proposed. When he accepted of the appointment the salary of his predecessor was doubled, and the rank of colonel was given him. Besides delivering his lectures as professor, he officiated also as inspector of the above corps, in the room of General Politika, who had retired, or been sent to his estates in the Ukraine.

The lectures which he gave were very much admired, and could not fail to be of the greatest use to his pupils. Few men understood so well the

theory and the practice of the arts they profess
to teach ; few had enjoyed the same opportunities
of seeing the mathematical rules of artillery and navigation carried into effect on so great a scale.   To
his own countrymen resident at Petersburgh, Mr
Robison was an object of no less affection than
admiration.

In 1773, the death of Dr Russell produced a
vacancy in the Natural Philosophy Chair of the
University of Edinburgh.   Principal Robertson,
who was ever so attentive to the welfare of the uni·
versity over which he presided, though not personally acquainted with Mr Robison, yet knowing
his character, had no doubt of recommending him
to the patrons of the university, who, on their
part, with no less disinterestedness, listened to
his recommendation, and Mr Robison was accordingly elected.   It is said, that when the news
of this appointment reached him, he at first hesitated about the acceptance of it, principally from the
fear of appearing insensible to the kindness and favour which he had experienced from the Russian
government.   The moment, too, when it was
known that this invitation had been given him, further offers of emolument and preferment were
made him by that government, of such a kind as it
was supposed he could not possibly resist.   At
length he determined, and no doubt wisely, however
splendid the prospects held out to him might be, to

accept of a situation that would fix him permanently in his native country. He therefore declined the offers of the Empress of Russia, and in June 1774 sailed from Cronstadt for Leith, followed, as one of those friends he left behind in Russia has expressed it, by the regrets, and accompanied by the warmest good wishes, not only of all who had shared in his friendship, but of all to whom he was known. The empress was so far from being offended with his determination, however much she wished to prevent it, that she settled a pension on him, accompanied with a request, that he would receive under his care two or three of the young cadets who were to be selected in succession.

Mr Robison was admitted at Edinburgh the 16th September 1774, and gave his first course of lectures in the winter following. The person to whom he succeeded had been very eminent and very useful in his profession. He possessed a great deal of ingenuity, and much knowledge, in all the branches of Physical Science. Without perhaps being very deeply versed in the higher parts of the mathematics, he had much more knowledge of them than is requisite for explaining the elements of Natural Philosophy. His views in the latter were sound, often original, and always explained with great clearness and simplicity. The mathematical and experimental parts were so happily combined,

that his lectures communicated not only an excellent view of the principles of the science, but much
practical knowledge concerning the means by which
those principles are embodied in matter, and made
palpable to sense.

Mr Robison, who now succeeded to this chair,
had also talents and acquirements of a very high
order. The scenes of active life in which he had
been early engaged, and in which he had seen the
great operations of the nautical and the military
art, had been followed, or accompanied, with much
study, so that a thorough knowledge of the principles, as well as the practice, of those arts, had been
acquired. His knowledge of the mathematics was
accurate and extensive, and included, what was at
that time rare in this country, a considerable familiarity with the discoveries and inventions of the
foreign mathematicians.

In the general outline of his course, he did not,
however, deviate materially from that which had
been sketched by his predecessors, except, I think,
in one point of arrangement, by which he passed
from Dynamics immediately to Physical Astronomy. The sciences of Mechanics, Hydrodynamics,
Astronomy, and Optics, together with Electricity
and Magnetism, were the subjects which his lectures embraced. These were given with great
fluency aud precision of language, and with the introduction of a good deal of mathematical demon-

stration. His manner was grave and dignified.
His views always ingenious, and comprehensive,
were full of information, and never more interesting and instructive than when they touched on the
history of science. His lectures, however, were
often complained of, as difficult and hard to be followed, and this did not, in my opinion, arise from
the depth of the mathematical demonstrations, as
was sometimes said, but rather from the rapidity of
his discourse, which was in general beyond the
rate at which accurate reasoning can be easily followed. The singular facility of his own apprehension, made him judge too favourably of the same
power in others. To understand his lectures completely, was, on account of the rapidity, and the
uniform flow of his discourse, not a very easy task,
even for men tolerably familiar with the subject.
On this account, his lectures were less popular than
might have been expected from such a combination
of rare talents as the author of them possessed.
This was assisted by the small number of experiments he introduced, and a view that he took of
Natural Philosophy which left but a very subordi-
nate place for them to occupy. An experiment, he
would very truly observe, does not establish a general proposition, and never can do more than prove
a particular fact. Hence, he inferred, or seemed
to infer, that they are of no great use in establishing the principles of science. This seems an er-

roneous view. An experiment does but prove a particular fact ; but by doing so in a great number of cases, it affords the means of discovering the general principle which is common to all these facts. Even a single experiment may be sufficient to prove a very general fact. When a guinea and a feather, let fall from the top of an exhausted receiver, descend to the bottom of it in the same time, it is very true that this only proves the fact of the equal acceleration of falling bodies in the case of the two substances just named ; but who doubts that the conclusion extends to all different degrees of weight, and that the uniform acceleration of falling bodies of every kind, may safely be inferred ?

A society for the cultivation of literature and science had existed in Edinburgh ever since the year 1739, when, by the advice, and under the direction of Mr Maclaurin, an association, formed some years before for the improvement of Medicine and Surgery, enlarged its plan, and assumed the name of the Philosophical Society. This society, which had at different times reckoned among its members some of the first men of whom this country can boast, had published three volumes of Memoirs, under the title of Physical and Literary Essays ; the last in 1756, from which time the society had languished, and its meetings had become less frequent. At the time I am now speaking of, it was beginning to revive, and its tendency

to do so was not diminished by the acquisition of
Mr Robison, who became a member of it soon after
his arrival.   It had often occurred, that a more regular form, and an incorporation by Royal Charter,
might give more steadiness and vigour to the exertions of this learned body.   In 1783, accordingly,
under the auspices of the late excellent Principal
of this University, a Royal Charter was obtained,
appointing certain persons named in it as a New
Society, which, as its first act, united to itself the
whole of the Philosophical.

Professor Robison, one of those named in the
original charter, was immediately appointed secretary, and continued to discharge the duties of that
office, till prevented by the state of his health several years after.

The first volume of the Transactions of this Society contains the first paper which Professor Robison submitted to the public, a " Determination
of the Orbit and the Motion of the Georgium
Sidus, directly from Observations," read in March
1786.   This planet had been observed by Dr Herschell on the 13th March 1781, and was the first
in the long list of discoveries by which that excellent observer has for so many years continued to
enrich the science of Astronomy.   Its great distance from the sun, and the slowness of its angular
motion, which last amounts to little more than four
degrees from one opposition to the next, made it

difficult to determine its orbit with tolerable accuracy, from an arch which did not yet exceed an
eighteenth part of the whole orbit. This was an
inconvenience which time would remedy ; but impatience to arrive even at such an approximation as
the facts known will afford, is natural in such cases,
and Professor Robison, as well as several other mathematicians, were not afraid to attempt the problem, even in this imperfect state of the data. It
is well known that the observations which best serve
the purpose of determining the orbit of a planet,
are those made at its oppositions to the sun, when
an observer in the Earth and in the Sun would refer the planet to the same point in the starry heavens, or when, in the language of astronomers, its
heliocentric and geocentric places coincide. Of
these oppositions in the case of this planet, there
were yet only four which had been actually observed. Dr Herschell had, however, discovered
the planet soon after the opposition of 1781 was
passed, and though of course that opposition was
not seen, yet from the observations that were made
so soon after, Professor Robison thought he could
deduce the time with sufficient accuracy. The
opposition of the winter 1786 he observed himself;
for though there was, unfortunately, no observatory
at Edinburgh, he endeavoured to supply that defect on the present occasion by a very simple appa
ratus, viz. a telescope on an equatorial stand, which

served to compare the right ascension and declination of the planet with those of some known stars which it happened to be near. His general solution of the problem is very deserving of praise; and though the method pursued is in its principle the same with all those which ever since the time of Kepler have been employed for finding the elements of a planetary orbit, it appears here in a very simple form, the construction being wholly geometrical, and easily understood. The elements, as he found them, are not very different from those that have since been determined from more numerous and more accurate observations.

When Dr Herschell first made known this most distant of the planets, many astronomers believed that they had discovered the source of those disturbances in our system, which had not yet been explained. Professor Robison was of this number; for he tells us, in the beginning of his paper, that he had long thought that the irregularities in the motion of Jupiter and Saturn, which had not been explained by the mutual gravitation of the known planets, were to be accounted for by the action of planets of considerable magnitude, beyond the orbit of Saturn. Subsequent inquiry, however, has not verified this conjecture; the irregularities of Jupiter and Saturn have since been fully explained, and are known to arise chiefly from their action on one another, a very small part only being owing to that of the Georgium Sidus, or of any of the other planets.

The next publication of Professor Robison was a paper in the second volume of the same Transactions, " On the Motion of Light, as affected by Refracting and Reflecting Substances, which are themselves in Motion." *

The phenomena of the aberration of the fixed stars are well known to depend on the velocity of the earth's motion combined with the velocity of light ; the quantity of the aberration, when all other things are given, being directly as the first, and inversely as the second. It is not, however, the general or the medium velocity with which light traverses space, but it is the particular velocity with which it traverses the tube of the telescope, that determines the quantity of this aberration. Were it possible, therefore, to increase or diminish that velocity, the aberration would be diminished in the first case, and increased in the second. But, according to the principles now generally received in optics, the velocity of light is increased, when it traverses a denser medium, or one in which the refraction is greater ; and, therefore, were the tube of a telescope to be filled with water instead of air, the aberration would be diminished. Professor Robison, and his friend Mr Wilson, Professor of

* Edinburgh Transactions, Vol. II. p. 83.

Astronomy at Glasgow, had speculated much on this subject, and made many attempts to obtain a water telescope, but, hitherto, without effect. A paper of Boscovich, on the same subject, seemed to suggest some new views, that might render the experiment more easy to be made. That philosopher maintained, that in ascertaining the effect of a water telescope on the motion of light, the observation of celestial objects might be dispensed with, and that of terrestrial substituted in its place. He argued, that while light moves with an uniform velocity, the telescope must be directed, not to the point of space which the object occupied when the particle was sent off which is entering the telescope, but to a point advanced before it by a space just equal to that which both the object and the observer have passed over in the time in which the particle has passed from the object to the eye. It is therefore directed exactly to the place which the object is in when the light from it enters the eye. If, therefore, the ray, on entering the telescope, is made to move faster than it did before, the telescope must not be inclined so much, and the apparent place of the object will fall behind its true place. If the ray is retarded on entering the water, the contrary must happen. Hence a number of very unexpected phenomena would result, affording, without having recourse to the heavenly bodies, a direct proof of the motion of the earth in its orbit, as well as a

resolution of the question, whether light is accelerated or retarded on passing from a rarer to a
denser medium. *

On this reasoning Professor Robison has very
well remarked, that it would be just, if the light,
on entering the water telescope, had only its velocity changed, and not its direction. But this is
not the case ; for the ray that is to go down the
axis of the telescope, is not perpendicular to the
surface of the fluid ; it makes an angle with it, depending on the aberration, and therefore in some
cases less by 20" than a right angle. On this account, the effect is not produced which Boscovich's
reasonings lead us to expect.

The sequel of the paper is also full of ingenious
remarks.

In December 1785, Mr Robison was attacked by
a severe disorder, which, with but few intervals of
relaxation, continued to afflict him to the end of his
life, and which, though borne with much resigna-
tion, and resisted with singular fortitude, could not
but at length impair both the vigour and the continuity of his exertions. The disorder seemed to be
situated between the urethra and the perineum. At
times it was accompanied with the severest pain, and
with violent spasms, which were easily excited. The
disease, however, was only known by the pain pro-

---

* Boscovich, Opera Math. Tom. II. Opusc. 3.

8

duced ; and never, by any visible or palpable symptom, gave information of its nature, as no change in
the parts which were the seat of it could ever be observed.   A complaint of this nature, it is evident,
must have less chance of being removed than any
other ; and it accordingly baffled the art of the
most skilful medical men, both in Edinburgh and
London.

Notwithstanding this state of suffering, his general health was not, for a long time, materially injur-
ed, nor the powers of his mind relaxed, so that he
continued to prosecute study with vigour and steadiness.   A malady which was both severe and chroni-
cal, admitted of no palliative so good as the comfort
of domestic society, which Mr Robison happily enjoyed, having married soon after he settled in Edinburgh.   The care and attention of Mrs Robison,
and the affectionate regards of his children, as they
grew up, were blessings to which, with all his habits of study and abstraction, he was ever perfectly
alive.

This indisposition did not prevent him from engaging, about this time, in a very laborious undertaking.   A work, with the title of Encyclopædia
Britannica, undertaken at Edinburgh several years
before this period, was now undergoing a third
edition, in which it was to advance from three to
eighteen volumes.   Twelve of these had been already published, under the direction of the original

editor, Mr Colin Macfarquhar, when, on his death, the task of continuing the work was committed to the care of the Reverend Dr Gleig ; and about the same time Professor Robison became a contributor to it. He was the first contributor who was professedly and really a man of science ; and from that time the Encyclopædia Britannica ceased to be a mere compilation. Dictionaries of arts and sciences, in this island, had hitherto been little else than compilations ; and, though in France, the co-operation of some of the most profound and enlightened men of the age had produced a work of great merit and celebrity, with us compositions of the same class had been committed to the hands of very inferior artists. The accession of Professor Robison was an event of great importance in the history of the above publication.

It was in the year 1793 that he began to write in this book, and it was at the article Optics, with him a very favourite science, that his labours commenced. From that time he continued to enrich the Encyclopædia with a variety of valuable treatises, till its completion in 1801.

The general merit of the articles thus composed makes it difficult to point out particulars. Those in which theoretical and practical knowledge are combined are of distinguished merit ; such are Seamanship, Telescope, Roof, Water-works, Resistance of Fluids, Running of Rivers. To these I

10

must add the articles Electricity and Magnetism in the Supplement, where the theories of Ӕpinus are laid down with great clearness and precision, as well as with very considerable improvements. In ascertaining the law of the electric attraction, his experiments were ingenious, as well as original, and afforded an approximation to the result which the great skill and the excellent apparatus of Coulomb have since exactly ascertained. In the Supplement is also contained a very full account of the Theory of Boscovich, a subject with which he was much delighted, and which he used to explain in his lectures with great spirit and elegance.

These articles, if collected, would form a quarto volume of more than a thousand pages. I am persuaded, that, when brought together, and arranged by themselves, they will make an acceptable present to the public ; and I have the satisfaction to state, that such a work is now preparing, under the direction of an editor, whose remarks or corrections cannot but add greatly to its value. Notwithstanding the merit which the separate articles possess, they are not entirely free from the faults incident to whatever is composed for a work already in the press. The condensation and arrangement, to which time is such an essential condition, even with men of the first talents, must be often wanting, in such circumstances ; and there are, accordingly, in the articles now referred to, a diffuseness, and sometimes a want of order, that may easily be corrected, without injuring the authenticity of the work.

Though the Encyclopædia employed Professor Robison very much during the whole of the seven years that it continued, he nevertheless found leisure for some researches of a very different nature. At the period of which I now speak, the French Revolution had arrested the attention, and excited the astonishment of all Europe; and the satisfaction with which the first efforts of a nation to assert its liberties, had been hailed from all quarters, was, by the crimes and excesses which followed, quickly converted into grief and indignation. A body was put in motion sufficient to crush whole nations under its weight; none had the power or the skill to direct its course; what movements it might communicate to other bodies, how far it would go, or in what quarter, it seemed impossible to foretell. The amazement became general; no man was so abstracted from the pursuits of the world, or so insulated by peculiarities of habit and situation, as not to feel the effects of this powerful concussion. All fixed their eyes on the extraordinary spectacle which France exhibited; where, if time is to be measured by the succession of events, a year was magnified into an age; and when in a few months one might behold more old institutions destroyed, and more new ones

projected or begun, than in all the ten centuries which had elapsed between Charlemagne and
the last of his successors ;—in a word, where
the ancient edifice, founded in the ages of bar-
barism with such apparent solidity, strengthened
and adorned in the progress of civilization with so
much skill and labour, was in one moment levelled
with the dust. A general state of alarm and
distrust was the effect of the convulsions which men
saw every where around them ; where the institutions held as sacred from their origin, or venerable
from their antiquity, and essential to the order
of society, were seen, not falling to pieces from natural decay, but blown up by the force of a sudden
and unforeseen explosion. From such a condition
of the world, jealousy and credulity could not fail
to arise. When danger is all around, every thing
is of course suspected ; and when the ordinary connection between causes and effects cannot be traced,
men have no means of distinguishing between
the probable and the improbable ; so that their opinions are dictated by their prejudices, their impressions, and their fears. Such, accordingly, was the
state into which men's minds were brought at this
extraordinary crisis ; and even in this country,
removed, as we were, from the danger, by so
strong a barrier of causes, both moral and physical,
the alarm was general and indiscriminate. The
progress of knowledge was supposed by many to be

the cause of the disorder ; panegyrics on ignorance
and prejudice were openly pronounced ; the serious
and the gay joined in declaiming against reason and
philosophy ; and all seemed to forget, that when
reason and philosophy have erred, it is by themselves
alone that their errors can be corrected.

The fears that had thus taken possession of men's
minds were often artificially increased. It was supposed that the general safety depended on the general alarm ; that the more the terror was extended,
the more would the object of it be resisted ;
and hence, doubtless, many felt it their interest,
and some considered it their duty, to magnify the
danger to which the public was exposed.

It is evident, that an inquiry into the causes
of the French Revolution, undertaken at a moment
of such agitation, was not likely to bear the review
of times of calm and sober reflection. It was
at this moment, however, and under the influence
of such impressions, that Mr Robison undertook to
explain the causes of that revolution. He was
deeply affected by the scenes that were passing before him. He possessed great sensibility, and his
mind, peculiarly alive to immediate impressions,
felt strongly the danger to which the social order
of every nation seemed now to be exposed. The
crimes which the name of Liberty had been employed to sanction, filled him with indignation, and the
contempt of religion, affected by many of the lead-

ers of the Revolution, wounded those sentiments of piety which he had uniformly cherished from his early youth.

In such circumstances, a mind accustomed to inquire into causes, as his had long been, could not abstain from the attempt to trace the sources of so extraordinary a succession of events. As to the circumstances which first led him, and led him, I think, so unhappily, to look for those sources in the institutions of Free Masonry, or in the combination of some German mystics, I have nothing satisfactory to offer. He was accustomed to refined and subtle speculations, and naturally entertained a partiality for theories that called into action the powers by which he was peculiarly distinguished.

In 1797, he published a book, entitled " Proofs of a Conspiracy against all the Religions and Governments of Europe." He supposes that this conspiracy originated in the lodges of the Free Masons, but that it first assumed a regular form in the hands of certain philosophic fanatics, distinguished in Germany by the name of *Illuminati ;* that after the suppression of this society by the authority of government, the spirit was kept alive by what was called the German Union ; that its principles gradually infected most of the philosophers of France and Germany, and lastly broke forth with full force in the French Revolution.

The history of *Illuminatism,* as it is called, forms the principal part of the work; and on a subject involved in great mystery, where all the evidence came through the hands of friends or of enemies, it was exceedingly difficult for one liv-- ing in a foreign country, and a stranger to the public opinion, to obtain accurate information. Accordingly, the events related, and the characters described, as proofs of the conspiracy, are of so extraordinary a nature, that it is difficult to persuade one's self that the original documents from which Mr Robison drew up his narrative were entitled to all the confidence which he reposed in them.

I do not mean to question the general fact, that there did exist in Germany a society having the vanity to assume the name just mentioned, and the presumption or the simplicity to believe that it could reform the world. In a land where the tendency to the romantic and the mysterious seems so general, that even philosophy and science have not escaped the infection, and in states where there is much that requires amendment, it is not wonderful if associations have been formed for redressing grievances, and reforming both religion and government. Some men, truly philanthropic, and others, merely profligate, may have joined in this combination; the former, very erroneously supposing, that the interests of truth and of mankind may be advanced by cabal and intrigue; and the latter, more wisely concluding, that these are engines well adapted to promote the dissemination of error, and the schemes of private aggrandisement.   An ex-Jesuit may have been the author of this plan, and whether he belonged to the former or the latter class, may have chosen for the model of the new arrangement, those institutions which he knew from experience to be well adapted for exercising a strong but secret influence in the direction of human affairs.

In all this there is nothing incredible; but the same, I think, cannot be asserted, when the particulars are examined in detail.   It is extremely difficult, as has already been remarked, for a foreigner, in such circumstances as Mr Robison's, to avoid delusion, or to determine between the different kinds of testimony of which he must make use. With me, who have no access to the original documents, and if I had, who have neither leisure nor inclination to examine them, an opinion can only be formed from the internal evidence, that is, from the nature of the facts, and the style in which they are recorded.   The style of the works from which Mr Robison composed his narrative, is not such as to inspire confidence; for, wherever it is quoted, it is that of an angry and inflated invective.   The facts themselves are altogether singular, arguing a depravity quite unexampled in all the votaries of *illumination*.   From the perusal of the whole, it is impossible not to conclude, that the alarm excited by the French Revolution had produced in Mr Robison a degree of credulity which was not natural to him. The suspicion with which he seems to view every person on the continent, to whom the name of a philosopher can be applied, and the terms of reproach and contempt to which, whether as individuals or as bodies, they are always subjected, make it evident that the narrative is not impartial, and that the author was prepared, in certain cases, to admit the slightest presumption as clear and irrefragable evidence. When, indeed, he speaks of such obscure men as composed the greater part of the supposed conspirators, we have no direct means of determining in what degree he has been misled. But when we see the same sort of suspicion and abuse directed against the best known and most justly celebrated characters of the age, we cannot but lament the prejudices which had taken possession of an understanding in other matters so acute and penetrating.

Among the men engaged in public affairs, of whom Europe boasted during the last century, there was perhaps none of a higher character than Turgot, who, to the abilities of a statesman, added the views of a philosopher; was a man singularly patriotic and disinterested, distinguished by the virtues both of public and private life, and having, indeed, no fault but that of being too good for the

times in which he lived. Yet Mr Robison has
charged this upright and humane minister with an
exercise of power, which would argue the most extreme depravity. He states, * that there existed
in Paris a combination under the direction of the
Wits and Philosophers, who used to meet at the
house of Baron D'Holbach, having for its object
the dissection of the brains of living children, purchased from poor parents, in order to discover the
principle of vitality. The police, he adds, interposed to put a stop to these bloody experiments,
but the authors of them were protected by the credit of Turgot.

All this is asserted on the authority, it should
seem, of some anonymous German publication. I
will not enter on the refutation of a calumny with
the fabrication of which Mr Robison is not charge-
able, though culpable, without doubt, for having allowed his writings to become the vehicle of it.
Truth and justice require this acknowledgment ;
and, in making it, I think that I am discharging a
duty both to Mr Robison and myself :—It is a
duty to Mr Robison, in as much as a concession
made by a friend, is better than one extorted by
an adversary ; it is a duty to myself, because I
should feel that I was doing wrong, were I even
by silence to acquiesce in a representation which I
believed to be so ill-founded and unjust.

---

* Proofs of a Conspiracy, &c. 4th Edit. Note, p. 584.

The Proofs of the Conspiracy, notwithstanding these imperfections, or perhaps on account of them, were extremely popular, and carried the name of the author into places where his high attainments in science had never gained admission for it. In the course of two years, the book underwent no less than four editions. It is a strong proof of the effect on the minds of men produced by the French Revolution ; and of the degree in which it engrossed their thoughts, that the history of a few obscure enthusiasts in Bavaria or Wirtemberg, when it became associated with that Revolution, was read in Britain with so much avidity and attention.

The defects of the evidence were concealed by the prejudices and apprehensions which were then so general. The people of this country were disposed to believe every thing unfavourable to the French nation, but particularly to the philosophers. All might not be equally culpable, but to discriminate between them was not thought of much importance, and it was the simplest, if not the fairest way, to divide the demerit equally among the whole. The rhapsodies of Barruel had already prepared the public for such impartial decisions, and had held up every man of genius and talents, from Montesquieu to Condorcet, as objects of hatred and execration.

But whatever opinion be formed of the facts related in the history of this conspiracy, it is certainly not in the visions of the German Illuminati, nor in the ceremonials of Free-masonry, that we are to seek for the causes of a revolution, which has shaken the civilized world from its foundations, and left behind it so many marks, which ages will be required to efface. There is a certain proportionality between causes and their effects, which we must expect to meet with in the moral no less than in the natural world ; ,in the operations of men, as well as in the motions of inanimate bodies. Whenever a great mass of mankind is brought to act together, it must be in consequence of an impulse communicated to the whole, not in consequence of a force that can act only on a few. A hermit or a saint might have preached a crusade to the Holy Land with all the eloquence which enthusiasm could inspire ; but if a spirit of fanaticism and of chivalry had not pervaded every individual in that age, they would never have led out the armies of Europe to combat before the walls of Jerusalem. Neither could the influence of a small number of religious or philosophic fanatics, sensibly accelerate or retard those powerful causes which prepared from afar the destruction of the French monarchy. When opposed to these causes, such influence was annihilated ; when co-operating with them, its effects were imperceptible. It was a force which could only follow those already in action ; it was like " dashing with

the oar to hasten the cataract," or, " waving with
a fan to give swiftness to the wind." *

It is, however, much easier to say what were not,
than what were, the causes of the French Revolution ; and in dissenting from Professor Robison, I
will only remark in general, that I believe the principal causes to be involved in this maxim, That a
certain relation between the degree of Knowledge
diffused through a nation, and the degree of Political Liberty enjoyed by it, is necessary to the stability of its government. The knowledge and information of the French people exceeded the measure that is consistent with the entire want of political liberty. The first great exigency of government, therefore, the first moment of a weak administration, could hardly fail to produce an attempt
to obtain possession of those rights, which, though
never enjoyed, can never be alienated. Such an
occasion actually occurred, and the revolution which
took place was entire and terrible. This also was
to be expected ; for there seems to be among political institutions, as among mechanical contrivances,
two kinds of equilibrium, which, though they appear very much alike in times of quiet, yet, in the
moment of agitation and difficulty, are discovered
to be very different from one another. The one is

---

* Ferguson's Essay on Civil Society, Part III. Sect. 4.

tottering and insecure, in so much that the smallest
departure from the exact balance leads to its total
subversion. The other is stable, so that even a
violent concussion only excites some vibrations
backward and forward, after which every thing
settles in its own place. Those governments in
which there is no political liberty, and where the
people have no influence, are all unavoidably in the
first of these predicaments : those in which there is
a broad basis of liberty, naturally belong to that in
which the balance re-establishes itself. The same
weight, that of the people, which in the first case
tends to overset the balance, tends in the second to
restore it : and hence, probably, the great difference between the result of the French Revolution,
and of the revolutions which formerly took place in
this country.

It will be happy for mankind, if they learn from
these disasters, the great lessons which they seem
so much calculated to enforce, and if, while the
people reflect on the danger of sudden innovation,
their rulers consider, that it is only by a gradual
reformation of abuses, and by extending, rather
than abridging, the liberties of the people, that a
remedy can be provided against similar convulsions.

But I return willingly from this digression, to
those branches of knowledge, where, in describing

what Mr Robison has done, the language of truth and of praise will never be found at variance with one another.

In autumn 1799, this country had the misfortune to lose one of its brightest ornaments, Dr Black, who had laid the foundation of the Pneumatic Chemistry, and discovered the principle of Latent Heat. The Doctor had published very little ; and his discoveries were more numerous than his writings. His lectures, however, had drawn much attention ; they presented the first philosophical views of chemical science ; they were remarkable for their perspicuity and elegance, and this, joined to the simplicity and gracefulness of manner in which they were delivered, made them universally admired. It was now proposed to publish these lectures ; but this required that they should be put into the hands of some one able to perform the part of an editor, and to prepare for the press the notes from which the Doctor used to read his lectures. The person naturally thought of was Mr Robison, one of Dr Black's oldest friends, and so well skilled in chemistry, that no one could be supposed to execute the work with more zeal or more intelligence. The task, however, was by no means easy. Dr Black, with a very large share of talent and genius, with the most correct taste and soundest judgment, with no habits that could dissipate his mind, or withdraw it from the pursuits of

science, was less ardent in research, and less stimu-
lated by the love of fame, than might have been
expected from such high endowments. A state of
health always delicate, and subject to be deranged
by slight accidents, was probably the cause of this
indifference.   Hence the small number of his writings, and his sudden stop in that career of discovery
on which he had entered with such brilliancy and
success.   Of much that he had done, the world had
never heard any thing, but from verbal communication to his pupils, and on the subject of latent
heat, no written document remained to ascertain to
him the property of that great discovery.   The
only means of repairing this loss, and counteracting
the injustice of the world, was the publication which
Professor Robison now undertook with so much
zeal, and executed with so much ability.   Dr Black
had used to read his lectures from notes, and these
often but very imperfect, and ranged in order by
marks or signs only known to himself.   The task
of editing them was therefore difficult, and required a great deal both of time and labour, but was at
last accomplished in a manner to give great satisfaction.   The truth, however, is, that the time was past
when this work would have met in the world with
the reception which it deserved.   Chemical theories
had of late undergone great changes, and the language of the science was entirely altered.   Dr
Black, on the subject of these changes, had cor-

responded with Lavoisier, and the mutual respect
of two great men for one another, was strongly
marked in the letters which passed between them.
The Doctor had acceded to the changes proposed
by the French chemist, and had even adopted the
new nomenclature ; but his notes had not undergone the alterations which were necessary to introduce it throughout.   It would now have been difficult to make those alterations ; and Mr Robison,
who was not favourable to the new chemistry, did
not conceive that, by making them, he was permanently serving the interest of his friend.   He
conceived, indeed, that there was unfairness in the
means employed by Lavoisier, for bringing Dr
Black to adopt the new system of chemistry, and
has thrown out some severe reflections on the conduct of the former, which appear to me to rest on
a very slight foundation.

It was quite natural for a man, convinced like
Lavoisier, of the importance of the improvements
which he had made in chemistry, to be desirous
that they should be received by the most celebrated Professor of that time,—by the very man, too,
whose discoveries had opened the way to those improvements.   His letters to Dr Black contain expressions of respect and esteem, which, I confess,
appear to me perfectly natural, and without any
thing like exaggeration or deceit.   Indeed, it is

not probable that M. Lavoisier, even if he could himself have submitted to flatter or cajole, could conceive that any good effect was to arise from doing so, or that there was any other way of inducing a grave, cautious, and profound philosopher, to adopt a certain system of opinions, but by convincing him of their truth. He had, with those who knew him, the character of a sincere man, very remote from any thing like art or affectation. We must, therefore, ascribe the view which Mr Robison took of this matter, to the same system of prejudices on which we have had already occasion to animadvert. Such, indeed, was the force of those prejudices, that he considered the Chemical Nomenclature, the new System of Measures, and the new Kalendar, as all three equally the contrivances of men, not so much interested for science, as for the superiority of their own nation. Now, whatever be said of the Kalendar, the project of uniform Weights and Measures is admitted to be an admirably contrived system, which Britain is now following at a great distance ; and the New Nomenclature of Chemistry to be a real scientific improvement, adopted all over Europe. Many of the radical words may depend on false theories, and may of course require to be changed ; but though the *matter* pass away, the *form* will remain ; the words of the language may perish, but the mould in which the language was cast will never be destroyed. *   The Lectures appeared in 1803.

The last of Mr Robison's works was one which he had long projected, though he now set about the completion and arrangement of it, for the first time. It was entitled, Elements of Mechanical Philosophy, being the substance of a Course of Lectures on that Science.   " Mechanical Philosophy" was, with him, a favourite expression ; it was understood as synonymous with Natural Philosophy, and included the same branches. The first volume, the only one he lived to finish, included Dynamics and Astronomy, and was published in 1804.   It is a work of great merit, and is accessible to those who have no more than an elementary knowledge of the mathematics. The short view of the phenomena prefixed to the Physical Astronomy is executed in a masterly manner. The same may be said, and perhaps even with more truth, of the Physical

---

* The high opinion which Mr Robison elsewhere expresses of Lavoisier is very remarkable. In his Astronomy, published a year after the Lectures, in stating Hooke's anticipation of the Principles of Gravitation, he concludes thus : " It is worthy of remark, that in this clear and candid and modest exposition of a rational theory, Hooke anticipated the discoveries of Newton, as he anticipated with equal distinctness and precision the discoveries of Lavoisier, a philosopher inferior perhaps only to Newton." (Elements of Mechanical Philosophy, p. 285.)

Astronomy itself; for there are very few of the elementary treatises on that branch of science which can be compared with it, either for the facility of the demonstration, or the comprehensiveness of the plan. The first part is meant to be popular and historical, and is so at the same time that it is philosophical and precise. The work is indeed highly estimable, and is entitled to much more success in the world than it has actually had.

We have already taken notice of Mr Robison's illness, with which he had been now afflicted for the long period of nineteen years. His sufferings, though not equal, had been often extremely severe. They had occasionally rendered him unable to discharge his duty in the College, and of late his friend, the Reverend Dr Thomas Macknight, had, with great kindness and ability, frequently supplied his place. Against such a continuance of ill health, with so little hopes of recovery as could be entertained for a long time past, hardly any mind could be expected to remain in full possession of activity and vigour. This is the more difficult, as the valuable medicine which alone in such cases can assuage pain, contributes itself at length to weaken the mind, and to destroy its energy. The combat which Mr Robison had maintained against these complicated evils, had indeed been wonderfully vigorous and successful, and the last of his works is quite worthy of his days of most perfect health and enjoyment.

The body could not resist so well as the mind. In the end of January 1805, he was suddenly seized with a severe illness, which put an end to his life in the course of forty-eight hours. There was a general disturbance of the system, which, without having the character of any defined disease, exhibited those symptoms of universal disorder which denote a breaking up of the constitution, and never fail to terminate fatally.

On reviewing the whole of his character, and the circumstances of his life, it is impossible not to see in him a man of extraordinary powers, who had enjoyed great opportunities for improvement, and had never failed to turn them to the best account. He possessed many accomplishments rarely to be met with in a scholar, or a man of science. He had great skill and taste in music, and was a performer on several instruments. He was an excellent draughtsman, and could make his pencil a valuable instrument either of record or invention. When a young man, he was gay, convivial, and facetious, and his *vers de société* flowed, I have been told, easily, and with great effect. His appearance and manner were in a high degree favourable and imposing ; his figure handsome, and his face expressive of talent, thought, gentleness, and good

temper.   When I had first the pleasure to become
acquainted with him, the youthful turn of his countenance and manners was beginning to give place
to the grave and serious cast, which he early assumed ; and certainly I have never met with any
one whose appearance and conversation were more
impressive than his were at that period.

Indeed, his powers of conversation were very extraordinary, and when exerted, never failed of producing a great effect.   An extensive and accurate
information of particular facts, and a facility of
combining them into general and original views,
were united in a degree of which I am persuaded
there have been few examples.   Accordingly, he
would go over the most difficult subjects, and
bring out the most profound remarks, with an
ease and readiness which was quite singular.   The
depth of his observations seemed to cost him nothing ; and when he said any thing particularly
striking, you never could discover any appearance
of the self-satisfaction so common on such occasions.
He was disposed to pass quite readily from one
subject to another ; the transition was a matter of
course, and he had perfectly, and apparently without seeking after it, that light and easy turn of
conversation, even on scientific and profound subjects, in which we of this island are charged by
our neighbours with being so extremely deficient.

The same facility, and the same general tone,

11

was to be seen in his lectures and his writings. He composed with singular facility and correctness, but was sometimes, when he had leisure to be so, very fastidious about his own compositions.

In the intercourse of life, he was benevolent, disinterested, and friendly, and of sincere and un-affected piety. In his interpretation of the con-duct of others, he was fair and liberal, while his mind retained its natural tone, and had not yield-ed to the alarms of the French Revolution, and to the bias which it produced.

His range in science was most extensive; he was familiar with the whole circle of the accurate sciences, and there was no part of them on which, if you heard him speak or lecture, you would not have pronounced it to be his *forte*, or a subject which he had studied with more than ordinary attention. In-deed, the rapidity with which his understanding went to work, and the extent of ground he seemed to have got over, while others were only preparing to enter on it, were the great features of his intel-lectual character. In these he has rarely been ex-ceeded. With such an assemblage of talents, with a mind so happily formed for science, one might have expected to find in his writings more of original investigation, more works of discovery and inven-tion. I must remark, however, that from the turn his speculations and compositions took, or rather re-ceived from circumstances, we are apt to overlook

what is new and original in a great part of them.
An article in a dictionary of science must contain a
system, and what is new becomes of course so mixed up with the old and the known, that it is not easily distinguished. Many of Mr Robison's articles
in the Encyclopædia Britannica are full of new and
original views, which will only strike those who
study them particularly, and have studied them
in other books. In *Seamanship*, for example, there
are many such remarks ; the fruit of that knowledge of principle which he combined with so much
experience and observation. *Carpentry, Roof,* and
many more, afford examples of the same kind.
The publication now under the management of Dr
Brewster, will place his scientific character higher
than it has ever been with any but those who were
personally acquainted with him.   With them, nothing can add to the esteem which they felt for his
talents and worth, or to the respect in which they
now hold his memory.

FINIS.

REVIEW

OF

MUDGE'S ACCOUNT OF THE

TRIGONOMETRICAL SURVEY OF

ENGLAND.

REVIEW

OF THE

TRIGONOMETRICAL SURVEY. *

THE work here announced is composed of papers
read at different periods in the Royal Society of
London, since the commencement of the Trigonometrical Survey in 1784, down to the present time.
As the interest excited by that survey created an
unusual demand for the volumes of the Philosophical Transactions in which the accounts of it were
contained, the publisher of this work thought he
would do a thing useful to science, and acceptable
to the public, by collecting all these accounts into
one.   In this he has had the assistance of the Royal
Society, and has been furnished with the plates
already engraved for the Transactions ; an indulgence of which he has made a very fair use, by
selling the book at a lower price than the elegance
of the work and the number of the plates might
have entitled him to demand.

---

* From the Edinburgh Review, Vol. V. (1805.)—ED.

The first volume of the Trigonometrical Survey was published in 1799; and it is only the second part of the second volume which, by its date, falls immediately under our notice; but we trust that the importance of a great national undertaking will justify the retrospect which we are about to take of the whole.

The idea of a Geometrical Survey, to be undertaken by Government, and executed at the public expence, first occurred on the suppression of the Rebellion in 1745, at the suggestion of the late Lieutenant-General Watson, at that time Deputy Quarter-master-general in North Britain. It fell to the late General Roy, who was then Assistant Quarter-master, to have a great share in the execution of this work; and the survey, which was at first meant to be confined to the Highlands, was extended to the low country, and made general for Scotland. Of the map produced from this survey, and which has remained in manuscript in the hands of Government, the General himself tells us, that though it answered the purpose for which it was intended, and is not without considerable merit, yet, the survey having been made with instruments of an inferior kind, and the sum annually allowed being very inadequate to so great a design, it is rather to be considered as a magnificent military sketch, than as an accurate map of a country.

At the conclusion of the peace of 1763, it came

under the consideration of Government to make a map of the whole island from actual survey, to which the map just mentioned was to be made subservient; the execution of the whole was to be committed to General, then Colonel Roy, whose experience, acquired in the Scottish survey, had been improved by as constant an exercise in the operations of practical geometry and astronomy as the duties of his profession would admit, and whom his love of such pursuits, and his indefatigable activity, pointed out as eminently qualified for this service.

Circumstances, however, which it is easy to conceive in general, but which it would be useless to know in detail, prevented any step from being taken toward the execution of this design, till after the peace of 1783, when a memoir drawn up by Cassini de Thury was presented to our Government by the French ambassador, setting forth the advantages that would accrue to geography and astronomy from carrying a series of triangles from Greenwich to Dunkirk, (to which place the meridian of Paris had already been extended,) so that the relative position of the two most celebrated observatories in the world might be ascertained by actual measurement.

This memoir having been communicated by the secretary of state (Mr Fox) to Sir Joseph Banks, and the plan proposed in it having received the approbation of the Royal Society, the execution of it was committed to General Roy, who was at that moment engaged in a survey of London and its environs, for the purpose of connecting together the different observatories in and about that metropolis ; a work which, with his usual ardour and activity, he had undertaken for his own amusement.

As a series of triangles was now to be extended from about Greenwich through Kent, and across the Channel to Calais and Dunkirk, the first thing to be done was to measure a base, from the length of which the lengths of the sides of all the triangles might be inferred. Such a line was accordingly traced out on Hounslow-heath, extending from a point near Hampton Poor-house, to a place called the King's Arbour, a distance of more than five miles, which was measured with the most scrupulous exactness. The description which General Roy has given of this measurement, deserves the attention of every one who is concerned in the operations of practical geometry, and who wishes to be made acquainted with the utmost resources of his art. He will perhaps see with surprise, that many of the things which he is accustomed to do with very little expence, either of time or of thought, require, when they are to be done with precision, no small proportion of both ; that to make two rods exactly of the same length, to place them in the same straight line, and to make the beginning of one coincide with the end of another, demand much skill and patience ; in a word, that the most common matter, when executed with extreme accuracy, becomes difficult ; and that science and art must combine to discover and to remove those minute obstacles, of which the bulk of mankind do not even suspect the existence.

The measurement of the base was first undertaken with deal rods of twenty feet in length ; but though these were made of the best seasoned timber, from an old mast cut up on purpose, though they were perfectly straight, and secured from bending in the most effectual manner, yet the changes in their lengths, produced by the moisture and dryness of the air, were so considerable, as to take away all confidence in the results deduced from them. Glass rods were therefore substituted in their room, consisting of straight tubes twenty feet in length, enclosed in wooden frames ; and these had the advantage of being susceptible of alteration only from heat or cold, according to laws which could be accurately ascertained. The base measured with these rods was found to be 27404.08 feet precisely, or 5.19 miles.

We refer for the particulars to the account itself, where General Roy has described the apparatus used, and the precautions taken to ensure the success of the operation. The detail, though minute, is interesting, and must be highly instructive to

those engaged in operations any way similar. We
would particularly recommend his description of the
deal rods, the method of laying off their lengths, of
the stands for supporting them, of the boning telescope, &c.

As the measurement of lines by a chain is, however, much more convenient and expeditious than
by any other means, it was thought desirable to
ascertain how far the accuracy of such a measurement could be depended on, and how near, in the
present instance, it might approximate to that by
the glass rods.   For the purpose of this experiment, Ramsden had prepared a chain of the very
best construction, made of hardened steel, one
hundred feet in length, and jointed somewhat like a
watch-chain.   General Roy having measured a
part of the base with this chain, and with the glass
rods at the same time, found that the results differed
by a quantity wholly inconsiderable.   Several years
afterwards, the whole base was measured with the
steel chain ; and the difference between that and
the measurement by the rods was found not to exceed two inches and three quarters of an inch, a
difference on the length of five miles that is plainly
of no account.   Hence it was inferred, that measurements made with such a chain as has just been
mentioned, and with due precautions, viz. stretching it always in the same degree, supporting it on
troughs laid horizontally, allowing for change of

temperature, &c. are as much to be relied on as those made in any other way whatsoever. This experiment, therefore, involved the determination of a material question with respect to the conduct of all future surveys. *

General Roy was assisted in these operations by Mr Isaac Dalby, a mathematician of eminence, and now Professor of the Mathematics in the Military College at High Wickham. A party of soldiers was also attached to the survey, for the purpose of doing such parts of the work as were merely laborious, and had a small encampment on the heath. The performance of this great experiment, for so it may very properly be called, could not fail to draw the attention of the men of science about London. The Master-General of the Ordnance, the President of the Royal Society, the Astronomer-Royal, and many other distinguished persons, frequently witnessed the skill and attention employed in conducting it. The mensuration of the base (including the repetition of it, and several collateral matters, as well as delays from bad weather) took up from the middle of June to the end of August 1784. The extremities of the base were then

---

* It may be proper to remark, however, that Legendre, after having considered the method of the steel chain, seems still to prefer that by *rods of metal*, because of the difficulty of giving the chain always a sufficient and uniform degree of tension.

marked by the centres of two wooden tubes, and have since been more permanently ascertained by the centres of two iron cannon sunk in the ground.

Experiments and observations, of the kind which we are now considering, seldom fail to benefit science, not only directly, but indirectly, by the collateral objects to which they lead. A pyrometer, constructed by Ramsden, for the purpose of ascertaining the expansion of solid bodies by heat, is probably the best instrument of the kind which has yet been made, and is one of the monuments of skill and genius that will long preserve the memory of that incomparable artist.

It was not till the summer of 1787 that the measurement was resumed, by actually extending a series of triangles from Greenwich to Dunkirk. For this purpose, signals were erected, in such conspicuous situations, and at such distances, as were judged convenient : the straight lines joining these points formed a set of triangles, the angles of which were measured by a theodolite which Ramsden had constructed, and which was carried, successively, to all the stations. In these triangles, therefore, which were so formed that one side was always common to two of them, all the angles became known ; and a side of one of them being also given, viz. the base on Hounslow-heath, the sides of all the rest could be found by trigonometrical computation. So also, the bearing of any one of the sides, in respect of the meridian, being known by observation, the bearings of all the rest with respect to the same meridian were determined.

The theodolite by which the angles of these triangles were measured, was superior to any thing that had ever been used in geodetical observations, and might be compared with the best instruments of astronomy. Ramsden had exerted himself to the utmost both in the design and execution of it : he had united in it the powers of a theodolite, a quadrant, and a transit instrument ; and had made it capable of measuring horizontal angles to fractions of a second. It was furnished with a telescope of a much higher magnifying power than had ever been before applied to observations purely terrestrial ; and by this superiority in its construction, even if it were the only one, we are persuaded that the surveys made with it are more accurate than any other. The French academicians, for example, who joined General Roy at Dover, as we shall see presently, employed in their measurement a very excellent instrument, a circle of repetition, of the kind invented by Borda ; and by taking the same angle several times over on different parts of the limb, they could diminish the error arising from the division of the instrument to an indefinite degree. But there was another error which they could not diminish, viz. that which arose from the small power of their glasses, and the consequent largeness of the real diameters of the objects which appeared to them as points or lines. Though their observations were therefore extremely good, and far exceeding any that had been made in France previous to the introduction of Borda's circle, they do not seem to equal those in the English survey. As England was later in undertaking works of this sort than France, and some other nations on the Continent, it seemed but just that she should aim at superior excellence ; and possessing, as Cassini says, the first artist in the world, it was not difficult for her to attain it.

The high power of the telescope just mentioned, obviated many of the difficulties concerning the signals employed to distinguish the precise point at each station that was to be intersected, from the rest. When the object to be intersected was not the spire of a steeple, a flag staff was commonly used ; but when the distance was great, or the weather not very favourable, *lights* were employed, and the observations were made in the night. These lights were either reverberatory lamps, or *white lights*, (so called from their extreme brillian-cy,) fired at a particular time previously agreed on. The signals made in this manner were visible at a great distance even in bad weather. Cassini says, that he hardly expects to be believed, when he tells, that he observed one from about Calais, which was fired on the opposite shore, about forty miles off, and in bad weather.

The precaution taken, of placing the great theo-dolite at all times with its centre exactly perpendicular to the point that was to be intersected from the other stations, deserves also to be mentioned. Though the allowance to be made for the distance of the instrument from the angular point is easily computed, yet it is difficult to avoid some error in doing so; and the frequent recurrence of such errors is a source of inaccuracy which it is much better to have entirely cut off.

From all these circumstances, added to others which we cannot here enumerate, the angles were generally observed with such accuracy, as to manifest the effect of the earth's sphericity, by giving the sum of the three angles of a triangle somewhat greater than 180°, and that even where the sides did not exceed 15 or 20 miles. This excess above 180° is produced by the plane of the instrument at the three angular points of the triangle not being parallel to itself, but perpendicular to three lines which meet (at least nearly) in one point, the centre of the earth. It is called the spherical excess; and it was in this survey that there came, for the first time, to be any question concerning the quantity of it, in each triangle. The instruments used in former surveys had never been accurate enough to bring a quantity so small as hardly ever to amount to 4″ in one triangle, to be an object of investigation. In the observations made for verifying the meridian of Paris about fifty years ago, the error in the three angles of a triangle often amounts to 20 or 30 seconds; and then, of course, no question could occur about a correction which cannot exceed the tenth part of that quantity. But in General Roy's observations, the error in the three angles of a triangle never reaches 3″; and, therefore, the spherical excess is of importance to be ascertained. In justice to the French academicians who co-operated with the General, it must be observed, that the angles taken by them with Borda's instrument, were accurate within 1″ or 2″ on each angle, so that they found equal reason for employing the spherical excess.

This remark is applicable to all the measurements made in France since the period we are here treating of. They have all been made with the *repeating* circle, and seem to have reached a very high degree of accuracy.

The introduction of a new element into trigonometrical computation is of great importance, and will probably be found to mark a precise era in all measurements relating to the figure of the earth. The spherical excess in any triangle has a given relation to the area of that triangle; for it is to 180° as that area is to the area of a great circle of the sphere; and it was from this theorem that General Roy deduced the rule which he has given for computing the spherical excess, independently of the angles themselves. For this purpose, the area of the triangle is to be estimated as if it were rectilineal; and it is sufficient to do this even by a very rude approximation, because it requires an area of about 75 square miles to produce a single second of spherical excess.

As it was necessary that the French geometers should unite with the English in carrying into full effect the plan which they themselves were the first to propose, three distinguished members of the Academy of Sciences, Cassini, Mechain, and Legendre, met General Roy and Dr Blagden at Dover, where measures were concerted for the corresponding observations to be made on the coasts of France and England. The French academicians were furnished by General Roy with white lights, to be fired on their side, while corresponding observations were made at Dover and Fairlight-Down, on the coast of England. The operations on both sides succeeded perfectly, notwithstanding that the weather was by no means favourable. The three academicians above named, having cross-ed the Channel again, after their observations were finished, repaired to London, and appear to have been highly gratified by the objects they saw, and the reception they met with in that metropolis. It

is painful to reflect, that this is the last amicable in-
terview which has taken place even among the men
of letters of the two countries; and that the hostile
armies of both nations are now encamped on the
very ground which was the theatre of these scientific operations.

Besides measuring all the angles in the triangles
that have been mentioned, it was necessary to
fix the bearing of some one of the sides of those
triangles in respect of the meridian. This was
done by observing the azimuth of the pole-star, relatively to the given line, at its greatest distance
from the meridian, both on the east and west sides.
This method of ascertaining the angle which any
line on the earth's surface makes with the meridian,
we apprehend to be greatly preferable, for expedition as well as accuracy, to any other that is known.
It cannot, however, be practised to advantage, but
with such an instrument as the great theodolite,
which answers for a transit, and carries a telescope
of power sufficient to render the pole-star visible
during the day. This, therefore, is one of the circumstances on account of which we think the British survey entitled to a preference above every
other.

That a check might not be wanting on any
errors that had crept into a work of such variety
and extent, General Roy caused a second base to
be measured on the flat ground of Rumney-Marsh,

which was not far distant from the southern extremity of the series of triangles. When the length
of this base, as actually measured, was compared
with that deduced by connecting it with the base
on Hounslow-heath, the two results were found to
differ only by twenty-eight inches, which must appear very inconsiderable, when we reflect that the
two lines are more than sixty miles asunder.
There was reason, nevertheless, to suspect that
this base of verification was not so correctly measured as that on Hounslow-heath.

The conclusions deduced from all these observations, as far as respects the relative position of the
observatories of Greenwich and Paris, are, first,
that the distance between their parallels of latitude is
963954 feet, = 182.567 miles, which corresponds
on the earth's surface to an arch of 2° 38' 26" in
the heavens, (the difference of latitude,) and therefore the length of a degree of the meridian in the
latitude 50° 10' comes out = 60843 fathoms =
69.14 miles.

Again, the perpendicular from the tower of
Dunkirk on the meridian of Greenwich is found to
be 547058 feet ; from which, subtracting 9080
feet, the distance of Dunkirk east of the meridian
of Paris, we have the perpendicular let fall from the
point in the meridian of Paris, which is in the parallel of Greenwich, on the meridian of this last

= 537978 feet = 101.89 miles. * The General
also having determined, from the length and
azimuth of one of the lines in the survey, (between
Botley-Hill and Goudhurst in Kent,) to how many
fathoms a degree of longitude in that parallel corresponds, has from thence deduced the difference
of longitude of Greenwich and Paris = $2° 19' 51''$,
or, in time, $9^m 19''.4$; which agrees with the conclusion which Dr Maskelyne had before drawn
from *data* purely astronomical. It must, however,
be observed, that Legendre deduces from the
same measurement a result considerably different,
and makes the difference of longitude of the two
observatories $9^m 21''$, (Mém. de l'Acad. 1788,)
which is $1\frac{1}{2}$ second in time greater than the preceding.

But though nothing certainly can exceed the

---

* Another result, not uninteresting, is the breadth of the
English Channel where it is narrowest. The line from the
Keep of Dover Castle to the station at Blancnez is 116660
feet = 22.095 miles. The South Foreland appears to be about
two miles nearer to Blancnez, if we measure on the map
which accompanies the survey. The least breadth of the
Channel, therefore, does not exceed twenty miles;—a narrow but a strong barrier,—one of those indelible lines which
nature has kindly traced out on the surface of the earth, to
resist the ambition, and preserve the independence of nations.

accuracy of General Roy's observations, we cannot bestow praise equally unqualified on the methods by which the results are deduced from them. The General has made use, as before mentioned, of the spherical excess, for the purpose of estimating the accuracy of his observations ; yet he has not derived, from the introduction of that new element, all the advantage which it is capable of affording. He has great merit in being the first to make use of it, though he did not perceive the whole of its importance. This was indeed first made known by a theorem of Legendre, in the Memoirs of the Academy of Sciences for 1787, from which it appears, that if each of the angles of a small spherical triangle be diminished by one-third of the spherical excess, the sines of the angles thus diminished will be very nearly proportional to the lengths of the sides themselves ; so that the computations with respect to such spherical triangles may be made by the rules of plane trigonometry. General Roy was probably unacquainted with this theorem, which is not of very easy investigation ; and though he has virtually employed it in part, because he always reduced the angles of his triangles to 180° before he used them in calculation, yet he derives no benefit from it in many of the cases where it is of the greatest importance. These are when two angles only of a triangle have been observed, and it is required to find the third angle ; or, again, in calcu-

lating the distances of the stations from the meridian, or the perpendicular to it, where the hypothenuse of a right-angled triangle is given, and one
of the oblique angles.   In such cases, an attention
to the above theorem will enable the calculator to
bring out a far more accurate conclusion than he
can otherwise obtain.   It must be confessed, that
the General has not calculated on this principle,
and that he has not taken as a substitute for it, the
reduction of the observed angles to the angles contained by the chords of the arches, (the method
that Major Mudge has adopted in the further prosecution of the trigonometrical survey ;) and that
therefore his calculations are deficient in accuracy,
at least in that extreme accuracy, which the correctness of the observations themselves entitles us
to expect.   His method is sufficiently correct for
any of the older measurements,—for those in Peru,
Lapland, France formerly, and indeed for all that
were made any where, till the great theodolite of
Ramsden, or the repeating circle of Borda, were introduced.   It in effect supposes the series of triangles to be laid out or extended on a plane surface,
and to this plane every thing is understood to be reduced.   This supposition is no doubt inaccurate ;
yet the inaccuracy is not considerable, and of no account at all, when the angles of a triangle have not
been observed within twenty or thirty seconds of the
truth.   But when the error of observation is re-

duced to less than a tenth part of that quantity, a more exact method of calculating must necessarily be pursued ; as the *calculus* should ever be so instituted as to preserve to the conclusions all the accuracy possessed by the *data* themselves.   No portion of this, however small, should be suffered to escape ; and since it is a quantity which the calculator cannot increase, he should be careful not to diminish it.

The slight degree of incorrectness, therefore, which we have remarked in General Roy's computations, would not deserve to be mentioned, if it were not for the excellence of his observations.   It is, besides, an imperfection which it is easy to remove : the part of the work which no one could amend, fortunately stands in need of no emendation.

It is not wonderful if these slight inaccuracies escaped General Roy.   The only principle on which they can be completely avoided, without a mode of calculating extremely long and laborious, is the theorem already mentioned ; a proposition by no means obvious, and drawn by the excellent mathematician who discovered it, from the recesses of the new geometry.   The General, possessing from his youth a decided turn for the mathematics, had bent the whole force of a very strong and vigorous understanding chiefly to the practical parts of that science, and those most immediately con-

nected with his profession ; but probably was not
much conversant with the branches which are more
theoretical and abstract.   A life spent in continual activity, and of which a large proportion had
been passed in the camp or the field, afforded no
leisure for such acquirements, and held out, even
to the mathematician, more interesting objects of
pursuit.   The duties of the field-engineer and the
quarter-master-general had particularly engaged his
attention ; and in every thing connected with them
he was profoundly versed.   He drew excellently,
and thoroughly understood the art of representing
the inequalities of ground with admirable distinctness, and great beauty of effect.   How perfectly
he was skilled in surveying, in the highest sense of
the word,—how conversant in the use of mathematical instruments, and in astronomical observation,
—it is unnecessary to state, after what has been
already said.   He was, besides, a most expert
and indefatigable calculator ; his acquaintance with
natural philosophy, too, was extensive and accurate ; and his paper on the measurement of heights
by the barometer, is a proof of his skill in conducting experimental inquiries, even when very remote
from the line of his ordinary pursuits.   General
Roy, it should be observed, had pursued this track
while the British army afforded few instances of
the same kind, either to encourage him by example, or rouse him by emulation, and when the con-

nection between the mathematical sciences and the military art was not so well perceived as it perhaps begins to be at the present moment.

The death of this excellent and accomplished officer, which took place soon after the period we now speak of, seemed for a long time to have put a stop to any design that might have been formed of extending the operations, already so happily begun, to the survey of the whole island: and here we must be permitted to remark, that the account given of the resuming of the survey is unsatisfactory and imperfect. After acknowledging the liberal assistance which the Duke of Richmond, as Master-general of the Ordnance, had given to every part of the preceding operations, (an acknowledgment which we believe to be most justly due,) it is said that a considerable time had elapsed without any apparent intention of renewing the survey, " when a casual opportunity presented itself (to the Duke of Richmond) of purchasing a very fine instrument, the workmanship of Ramsden, of a construction similar to that which was used by General Roy, but with some improvements; as also two steel chains, of one hundred feet each, made by the same incomparable artist."

Are we then to suppose that a great and national object was in danger of being dropt, or indefinitely delayed, but for a fortunate and unforeseen accident ? Did not the instrument which General

Roy had used still remain in the possession of the Royal Society ? and if the work was now to be prosecuted, not under the immediate direction of that society, but of the Board of Ordnance, can we suppose that, on that account, the use of it would have been withheld ? This is the less probable, that it has since been actually put into the hands of Major Mudge, and is at present employed by him in the survey. But be this as it may, the purchase of the new theodolite by the Duke of Richmond was indeed purely accidental ; for it had been made, if we are not misinformed, by order of the East India Company, for the purpose of surveying their possessions in the East ; and Ramsden, in the construction of it, had exerted that increased ingenuity and attention with which the presence of a great and new object used always to inspire him. In the end, some misunderstanding arose ; and a fit of ill-humour, or of ill-timed economy, induced the sovereigns of India to refuse an instrument which could do nothing to enlarge their dominions, though in skilful hands it might have done much to render them more secure. The Duke of Richmond was a better judge of its value ; and has rendered it no less useful to the public, than if it had followed its original destination.

In 1791, Captain Mudge of the Royal Artillery, and Mr Dalby, who has been already mentioned, both well qualified for the work they were to undertake, had the care of the trigonometrical survey committed to them, and received their instructions from the Master-General of the Ordnance. They began by the remeasurement of the base on Hounslow-heath with the new steel chain, (of the same nature with the former, but somewhat improved,) and found its length, as before stated, two inches and three quarters greater than when measured with the glass rods. The chain was here reduced, as it had been before, to the temperature of 62°, Captain Mudge having previously ascertained, by a series of experiments made with the chain extended at its full length, and stretched with a considerable weight, that it lengthened 0.0075 of an inch for one degree of heat, on Fahrenheit's thermometer; which agrees well with General Roy's determination of the same by means of the pyrometer.

As a series of triangles was now to be carried from Hounslow-heath to the coast of Kent and Sussex, and from thence westward to the Land's-End, it was thought right to measure another base of verification on Salisbury-Plain. This was done with all the precautions used in the former measurements; the length of the line was found to be 36574.4 feet; and when this was connected by a series of triangles with the base on Hounslow-heath, and its length deduced from this last by trigonometrical calculation, it did not differ by more than an inch from the actual measurement as here

set down. This singular coincidence was a sufficient proof of the accuracy with which the two
bases and the angles of the connecting triangles had
been measured.

One of the principal objects now in view was of
importance, both in general geography, and in the
topography of England. This was the measurement
of a degree of a circle perpendicular to the meridian, for which two stations on the coast of the
Channel, Beachy-Head in Sussex, and Dunnose in
the Isle of Wight, afforded a good opportunity, being visible from one another in fine weather, though
more than sixty-four miles distant, and the line between them being not far from the direction of
east and west. The distance between the two stations just mentioned, as deduced from a mean of
four different series of triangles, is 339397 feet,
(=64.28 miles;) and it is remarkable, that the ex-
tremes of these four determinations do not, even in
so long a line, differ more than seven feet from one
another. But coincidences of this sort are frequent
in the trigonometrical survey, and prove how much
more good instruments, used by skilful and attentive observers, are capable of performing, than the
most sanguine theorist could have ever ventured to
foretell. In two distances that were deduced from
sets of triangles, the one measured by General Roy
in 1787, the other by Major Mudge in 1794, one
of 24.133 miles, and the other 38.688, the two

measures agree within a foot as to the first distance, and sixteen inches as to the second. Such
an agreement, where the observers and the instruments were both different, where the lines measured were of such extent, and deduced from such a
variety of *data*, is probably without any other example. We believe it is quite unnecessary to add,
that these deductions are all made in the fairest and
most unexceptionable manner, without any means
being taken, purposely to bring about a coincidence
that would not have otherwise taken place.

Besides the determination of the distance from
Dunnose to Beachy-Head, the azimuth or bearing
of the line between them, with respect to the meridian, was carefully observed, by means of the
pole-star, after the manner practised by General
Roy. From these observations Major Mudge has
drawn the following conclusions. At Beachy-
Head, in latitude 50° 44′ 24″, the degree of longitude, measured on the above parallel, is 38718 fathoms; the degree of a circle perpendicular to the
meridian 611832; the degree of the meridian itself in that latitude being taken at 60851 fathoms,
as deduced from General Roy's measurement.
From the lengths of these degrees of the meridian,
and of the circle perpendicular to it, it follows,
that if the earth be an ellipsoid, the diameter of the
equator is to the polar axis as 149 to 148, which
makes the inequality between these two lines more

than twice as great as it appears to be by taking the
most probable average, deduced from all the observations that have been made in different latitudes.
What reason can be assigned for this peculiarity in
the physical constitution of our island, seems impossible at present to explain, though the continu-
tion of the Trigonometrical Survey may be expected to throw some light on it.   Local causes may
perhaps affect the direction of gravity in the south
of England, and may make that country appear to
be a portion of a smaller and more oblate spheroid,
than agrees with the general configuration of the
earth's surface.   Or perhaps, too, as many have
imagined, and as Major Mudge seems disposed to
think, the figure of the earth is not a solid formed
by the revolution of an ellipsis on its axis ; and the
agreement or disagreement of the measures of degrees with one another, is not to be judged of by
their agreement or disagreement with this hypothesis.   To attempt to judge of them in that manner, may be offering violence to nature, and may
be only trying to reconcile her phenomena with
our conjectural or arbitrary theories.

From the prosecution of the Trigonometrical
Survey we may expect a solution of these questions:
the unexpectedness of the results makes the work
more valuable to science ; and as we are sure that
the observations are accurate, the less they agree
with our preconceived opinions, the more interest-

ing do they become, and the more likely are they to furnish important information.

It has already been observed, that Major Mudge deduced the results of his observations on a principle more accurate than General Roy, by reducing every angle measured with the theodolite to a plane passing through the three angular points of the triangle, and thus computing the chords instead of the arches themselves.

" As the lengths," says he, " of small arcs and their chords are nearly the same, it is evident that the calculations might be performed sufficiently near the truth in any extent of a series of triangles, by plane trigonometry, if the angles formed by the chords could be determined pretty exact. We have endeavoured to adopt this method in computing the sides of the principal triangles, in order to avoid an arbitrary correction of the observed angles, as well as that of reducing the whole extent of the triangles to a flat, which evidently would introduce erroneous results, and these in proportion as the series of triangles extended." Vol. I. p. 271.

Now, concerning the method of calculation here referred to, we must observe, that though it is certainly much preferable to that which supposes the triangles to be all spread out over one flat surface, and is not liable to any considerable inaccuracy, yet is it much more complex and operose than one which we have already pointed out as being de-

rived from Legendre's theorem, that the arches
which form the sides of small spherical triangles,
are proportional to the sines of the opposite angles,
when each of these angles is diminished by the
third part of the *spherical excess*. By means of
this proposition, the sides themselves may be directly computed, and the investigation of the chords
avoided as unnecessary. From this it also appears,
that the correction made on the observed angles,
by taking from each of them one-third of the spherical excess, can be no longer viewed as an *arbitrary correction*, but as a legitimate and necessary
inference from a geometrical theorem extremely
curious in itself, as it marks the *continuity* of plane
and spherical triangles ; and extremely valuable, as
it leads to the most accurate and simple rules of
calculation.   Though it is a theorem that, in strictness, is only an approximation to the truth, yet its
accuracy, in all such cases as can come under consideration in a survey of any portion of the earth's
surface, may be safely relied on, the quantities
which it rejects being then really evanescent.

   In reducing the observed angles in the manner
of Major Mudge, (which is also that of Delambre,) there is always the inconvenience of an operose
and unnecessary *calculus ;* and, in certain cases,
such as the computation of the distances of the stations from the meridian or the perpendicular to it,
it is not quite evident but that inaccuracies of some

consequence may be introduced. If, on the other hand, in the solution of this last problem, since the hypothenuse and one of the oblique angles of a right-angled triangle are given to find the sides; if we first calculate the spherical excess, and proceed to find the other oblique angle by making the sum of it and the given angle=90°+that excess; if we then subtract one-third of the spherical excess from each of these oblique angles, and, with the angles so corrected, compute the sides by the rules of plane trigonometry; we shall obtain them with great ease, and with all the precision that the problem admits of. It would seem, then, that this last method of calculation is greatly preferable to the former.

We are perfectly aware of the caution with which theoretical men, sitting quietly in their closets, should offer advice to those who add the practice of art to the speculation of science ; who sacrifice ease and comfort to literary pursuits, and earn their reputation not merely by deep study, but by the sweat of the brow and the labour of the hands. We feel the full force of this maxim, and are writing at the present moment under the strongest impression of its truth. Yet, when we venture to recommend the method of calculation just described, as fit to be employed in the trigonometrical survey, we are not much afraid that the person best able, and most interested to judge correctly on that subject, after making the trial, will be inclined to censure our rashness.

The length to which these remarks have already extended, forces us to pass over the operations of several of the subsequent years when the survey was carried westward as far as the Land's-End, and again eastward to the remaining part of Kent, to Essex and the interior of the country, Oxfordshire, &c. Though these accounts are interesting from the importance of the places surveyed, and particularly from the drawing of four different meridians, and the determination of their difference of longitude, we shall pass to the consideration of the last thing performed in the survey, which is the measure of an arch of the meridian between Dunnose in the Isle of Wight, and Clifton near Doncaster. For the purpose of this measurement, Major Mudge was furnished with a new instrument, of the workmanship of Ramsden, viz. a zenith sector for the celestial observations, which were now required to be made with the greatest possible exactness. Though several instruments of this kind have been constructed by former artists, and many of them excellent, yet that which we have just mentioned seems greatly superior to them all. In it, the defects of former constructions are obviated, and many new improvements introduced. Among these must be reckoned the method of suspending the instrument, of bringing it into the vertical plane, of turning it to face opposite ways, and, most of all, the contrivance for adjusting the plumb-line perpendicularly over the centre of the sector, in which the skill of the optician and the mechanist are eminently combined. The instrument is supported by a strong pyramidal frame ; the telescope is an achromatic, eight feet long ; the radius of the sector is nearly the same ; and the angles measured on its limb may be read off to decimals of a second. The whole is a masterpiece of original design and skilful execution ; and, to its intrinsic value, adds that of being almost the last work of an artist, who was never equalled by those who went before him, and will not soon be surpassed by those who shall come after.

The meridian of Dunnose, in the Isle of Wight, was pitched on by Major Mudge to be extended northward, as affording a better opportunity than any other of measuring on it a large arch, free, as far as could be foreseen, from the action of any disturbing force. The meridians further to the west, though they might be produced to a greater length before they reached the sea, entered sooner into a mountainous, or at least a hilly country, where the direction of gravity must be affected by the inequalities of the surface. The meridian of Dunnose, traversing the plains of Hampshire, Berkshire, &c. and so on to Yorkshire, intersects the

sea-coast near the mouth of the Tees, without having passed over any high land, except on the con-
fines of Northamptonshire and Warwickshire, near
the sources of the Nen and the Avon, where the
ground rises to the height of 800 or 900 feet above
the level of the sea, with a gradual slope both to
the south and to the north. The part of this line
which has been actually measured, extends from
Dunnose to Clifton, not far from Doncaster, a distance of more than 196 miles, the length of which
was determined by a series of triangles carried from
one end of it to the other, like those that have been
already mentioned. The sides of these triangles
were deduced from the bases on Hounslow-heath
and Salisbury-plain, as in other parts of the survey ;
but, for the greater security, another base of about
five miles was measured on Misterton Car, near the
northern extremity of the chain of triangles ; and
this was done with the same precautions which
have been before enumerated. The latitude of
Dunnose and Clifton were then accurately determined by the sector, from stars near the zenith,
observed in their passage over the meridian. The
instrument was also carried to Ardbury-hill, (near
Daventry in Northamptonshire,) not far from the
middle of the line, the latitude of which was also
exactly determined. Besides this, as Blenheim was
not far distant from this meridian, and as its parallel of latitude had been accurately determined by

1

the Duke of Marlborough, from a series of observations made with the best instruments, the intersection of this parallel with the meridian could be fixed with accuracy, and gave of course another subdivision, both of the geodetical distance and the celestial arch. The same was done with respect to the parallel of Greenwich ; and thus, besides the total length of the meridian line from Dunnose to Clifton, there were given three intermediate points in that line, with the distances between them, and also the amplitudes of the corresponding arches in the heavens. From the comparison of all these determinations, some curious and unexpected conclusions have been deduced.

1. The entire length of the meridian line, from Dunnose to Clifton, is 1036337 feet, or 196.29 miles ; the latitude of Dunnose being 50° 37′ 8″.21, and the arch between its zenith and that of Clifton 2° 50′ 23″.38. Hence, the length of a degree in the middle between these places, or in lat. 52° 2′ 20″, is 60820 fathoms.

2. In the same way, by computing the length of a degree for the middle latitude between Ardbury-Hill and Dunnose, viz. 51° 35′ 18″, it is found equal to 60864 fathoms. This is 44 fathoms greater than the former, though, being about 27′ more to the south, it ought, according to every notion of the earth's oblate figure, to be several fathoms less.

3. Comparing, in like manner, the distance between Ardbury-Hill and Clifton with the arch intercepted by their zeniths, the degree in the latitude
of 52' 50' 30" is 60766 fathoms ; which is less than
either of the former, though, by being a good deal
further to the north, it ought to be considerably
greater.

4. The intersections of the parallels of Blenheim
and Greenwich with this meridian, give results of
the same kind, all tending to show that the degrees
diminish on going from the south to the north,
though not regularly, nor according to any law yet
known.   These inconsistencies are very striking,
when it is considered that, on the supposition of
the earth being compressed at the poles, the degrees
of the meridian must go on increasing as we proceed
northward, and in our latitudes nearly at a uniform
rate ; each degree exceeding that immediately to
the south of it by about twenty fathoms, according
to the theories that make the earth's oblateness the
greatest; and about ten, according to those which
make it the least.

To whatever cause these irregularities are to be
attributed, it cannot be, we are well convinced, to
the inaccuracy of the observations.   The probable
limits of such inaccuracy are considered by Major
Mudge himself ; and though he estimates them
as very small, yet, by any one who has carefully
studied the observations themselves, and remarked

their closeness to one another, he will not be thought to have diminished them more than the circumstances authorized him to do.

He states it as his opinion, after a re-examination of all his *data*, that the extreme of the error in the measurement of the whole distance, though nearly 197 miles, is not more than 100 feet, answering to about 1″ of a degree, and that the probable error does not exceed one half of that quantity. In the determination of the celestial arch he does not state so precisely his estimation of the error, or the limits within which it is contained ; but, taking in the multitude and the agreement of the observations, we should imagine that at any one station it can hardly amount to a second.   It is therefore to the action of some external cause, affecting the direction of the plumb-line, that the irregularities above stated are to be ascribed.   " I am disposed," says the Major, " to believe that the plumb-line was drawn towards the south, from the action of matter both at the northern extremity of the arch and at Ardbury-hill ; but more particularly at the first-mentioned station."—" The general tenor of the observations seems to prove that the plumb-line of the sector has been drawn toward the south at all the stations, and that by attractive forces, which increase as we proceed northward." From what physical cause this attraction proceeds,—from what circumstance in the structure or formation of the island, he does not offer any conjecture, neither shall we presume to do so. The continuation of the meridian to the north will probably throw some further light on this interesting subject.

It is, however, material to be observed, that when the degrees are irregular, as they appear to be here, the magnitude of the middle degree between two given latitudes is not rightly found, by dividing the terrestrial distance by the celestial arch. This process is only correct, on the supposition that the degrees increase or decrease in arithmetical progression, or at an uniform rate : if they vary according to any other law whatever, the degree found by the above operation will not be the degree in the middle point of the arch. This caution is necessary to be attended to, if we would deduce from the observations no more than what necessarily follows from them. It may be further remarked, that in the doubt we are in about the figure of the earth, whether it be a solid of revolution, and whether different meridians may not be unequal and dissimilar curves, it may be questioned whether the places on one meridian can be safely reduced to another, by the supposed intersections of their parallels with this last; and whether, by supposing such reduction, as when the observatories of Greenwich and Blenheim are placed on the meridian of Dunnose, we do not complicate the question unnecessarily by the introduction of a new and unknown element. The distances of these places to the eastward of the meridian being but small, it is indeed probable that, in the present case, any error introduced by them must be very inconsiderable ; but it is at least right to be apprised of the possibility of its existence.

Though we have no doubt that irregularities, which are at present so difficult to be accounted for, will, by the prosecution of the subject, become perfectly explained, we confess that we have felt some disappointment on reflecting, that hitherto the more that has been done to ascertain the figure of the earth by the measurement of degrees, the less satisfactory in some respects, has our knowledge of it become. The more microscopically we have observed, the more irregularities we have discovered ; and in the last experiment, which may be justly reckoned the best, what is accounted the natural order of things, has been almost completely inverted. All this, however, is only a motive for continuing the research ; which, if prosecuted with skill and perseverance, must ultimately lead to the knowledge of the truth. The time was, when the planetary motions were involved in the same confusion, and seemed the more unaccountable and perplexed, the more carefully they were studied. We may hope for the same issue in both cases, and that the figure of the earth will one day be as perfectly known as the orbit of a planet.

In such circumstances, we may congratulate the public, or those, at least, who are interested in the progress of science, on the continuance of the Trigonometrical Survey of England, notwithstanding the long and expensive wars in which the country has been engaged since the commencement of it. The expense of the Survey, indeed, is of little moment, compared with the object to be attained by it; but, in all times of difficulty, and in all plans of economy, the indulgences most intellectual and scientific are the first things to be sacrificed. It is to the credit of Government that it has been so far otherwise in the present instance.

A reflection, naturally called forth by the contemplation of so much accuracy as is displayed in the whole of the work now under our review, is, how much slower the mathematical arts have advanced than the mathematical sciences. Though the former were no doubt the first to start in the progress of improvement, they appear to have fallen behind, almost from the first outset. The rude manner in which Archimedes measured the apparent diameter of the sun is well known; and while that great geometer was investigating the properties of the sphere and cylinder with an acuteness and depth that have been the admiration of all succeeding ages, he was resolving one of the simplest problems of practical astronomy, in a more inaccurate manner than would be suffered in an ordinary seaman of modern times. When the great problem of measuring the circumference of the earth was first thought of, the principle upon which the solution was attempted was perfectly scientific, and the same, in fact, with that which we have just been considering; but the execution, though in the hands of able mathematicians, was *slovenly* and inaccurate in the extreme. The academicians of modern Europe have traversed the globe, from the equator to the polar circle, in order to resolve this great problem, and are still labouring hard, as we have seen, to give perfect accuracy to their conclusions. The academicians of Greece and Egypt put themselves to no such inconveniency. Eratosthenes, when he engaged in the inquiry, never quitted his observatory; but having measured the sun's solstitial elevation at Alexandria, where he lived, he took for granted, on report, that on the same day the sun was in the zenith of Syené, being seen there from the bottom of a deep well. He also assumed on no better authority, the distance and bearing of the two places, and, with such *data*, was not ashamed to say that he had computed the circumference of the earth.

At a much later period, our countryman Norwood set about determining the circumference

11

of the earth, with an accuracy as much superior to that of the Greek geometer, as it was inferior to that which has been the subject of the preceding remarks.   Having determined the latitudes of London and York by observation, he travelled from the one place to the other, measuring along the high-road with a chain, and taking the bearings with a compass.   He was well satisfied with the accuracy of his work.   " When I measured not," says he, " *I paced ;* and I believe the experiment has come within a *scantling* of the truth."

It is curious to compare these early essays of practical geometry with the perfection to which its operations have now reached, and to consider, that while the artist had made so little progress, the theorist had reached some of the sublimest heights of mathematical speculation ; that the latter had found out the area of the circle, and calculated its circumference to more than a hundred places of decimals, when the former could hardly divide an arch into minutes of a degree ; and that many excellent treatises had been written on the properties of curve lines, before a straight line of considerable length, had ever been carefully drawn, or exactly measured, on the surface of the earth.

FINIS.

REVIEW

OF

MECHAIN ET DELAMBRE;

BASE DU SYSTEME METRIQUE DECIMAL.

REVIEW

OF

MECHAIN ET DELAMBRE. *

I T is remarkable, that some of the clearest of our
ideas are incapable of being accurately expressed by
means of language, or of any arbitrary symbols
whatsoever. This happens with respect to certain
ideas of quantity ; while, with respect to others not
more clear or definite, the contrary takes place.
Of the magnitude of a line, for instance, no precise
notion can be conveyed in words from one man to
another, except by comparing it with a line already
known to them both ; and if a such a standard of
comparison is wanting, the ordinary means of information fail entirely, and there is no resource but in
the actual exhibition of the line itself. It is quite
otherwise, again, where either the ratio or the angular position of magnitudes are concerned : these
can be fully explained by verbal communication,
and never require the production of the objects

---

* From the Edinburgh Review, Vol. IX. (1807.)—Ed.

themselves. We know what a Greek geometer meant by a right angle, or by an angle of one degree, just as well as if we had before our eyes a circle divided by some artist of Athens or Alexandria. We understand, too, what he means when he speaks of the ratio of two to one, or of the ratio of the diagonal of a square to its side ; but if he specifies some individual length, of a foot for example, a spithame, or a stadium, we comprehend nothing of the matter, unless he has made a reference to some common standard, that is, to some magnitude which remains the same now as when he wrote.

So also when Eratosthenes tells us that the distance between Alexandria and Syené subtends, at the earth's centre, an angle, which is the fiftieth part of four right angles, we are at no loss to comprehend what is meant ; but when he says that the distance between the two places is 5000 stadia, we receive no accurate information ; and much critical discussion has been required to extract even a very uncertain meaning from his words.

This imperfection of language is founded in the nature of things, and is impossible to be removed. The inconveniences arising from it have been felt not only by the learned and scientific, but by all who have been concerned about measuring, weighing, or computing, even in the most imperfect state of the arts. In the measures of every country, we

may perceive attempts to obviate the difficulties
which have just been mentioned, and must feel
some interest in remarking the expedients adopted
for that purpose by rude and unenlightened men.
The *foot* which we recognize among the measures
of almost all nations, was taken from the standard
of the human foot, and varies, accordingly, within
limits of no very considerable extent.   Other
standards, supposed more precise, were sometimes
had recourse to.   Among agricultural nations, the
inch has been determined by the length of three
barley corns; and to the equestrian tribes of Arabia,
the breadth of a certain number of hairs from a
horse's tail afforded a standard of the same kind.
In weights, a drop of water appears to have been
regarded as a unit, according to some methods of
reckoning; and, according to others, a grain of
wheat stood for the weight which still takes its
name from that origin.   Some authors would have
us believe that the ancients, in their attempts to
form a standard measure, had proceeded very far
beyond these rude essays.   Paucton, in his *Métro-
logie*, will have it, that the circumference, or the
diameter of the earth, was the standard to which
they referred in their measures of length.   Bailly
has supported the same opinion, with the ingenuity
and learning displayed in all his speculations; and
he endeavours to prove, that the stadium was always taken for an aliquot part of the earth's cir-

cumference, that part being different with different
nations, and with different authors. No ingenuity,
however, can render this supposition probable.

The ancients had no means of determining, with
any tolerable precision, the magnitude of that great
unit to which their measures are supposed to refer. Besides, if such a reference had been intended, it could not surely have been unknown to themselves; yet we are well assured, that neither Aristotle, nor Possidonius, nor Pliny, nor any other ancient author who lays down the dimensions of the
globe, conceived that the difference between him
and other writers was only apparent, or that he
agreed with them about the magnitude of the earth,
and differed only about the length of the measure in
which he chose that its dimensions should be expressed. The first attempt at fixing such a standard of
measure as should be accurate, and universal, both
as to place and time, is due to the inventive genius
of the celebrated Huygens. That philosopher demonstrated that the times of vibrations of pendulums depend on their length only; and, whatever
be their structure, that a certain point may be
found, which in pendulums that vibrate in the same
time, is constantly at the same distance from the
centre of suspension. Hence he conceived that the
pendulum might afford a standard, or unit, for
measures of length; and though a correction would
be necessary, because the intensity of gravitation

10

was not the same in all latitudes, he believed that science furnished the means of determining this correction with sufficient accuracy. Picard laid hold of the same notion, and Cassini, in his book De la Grandeur de la Terre, proposed another unit, taken also from nature, though not so easily obtained, viz. the six thousandth part of a minute of a degree of a great circle of the earth. A similar idea had even earlier occurred to Mouton. No attempt, however, was made to raise, upon any of these standards, a regular system of measures, adapted either to the purposes of science or of ordinary life. Among the measures and weights that actually prevailed throughout Europe, the utmost confusion and perplexity continued to take place. In each sort of measure units of different magnitudes were admitted. These were inaccurately divided, and variously reckoned, to the disgrace of the economical arrangements of every country where they were found. The inconveniences which arose from thence were generally felt, and complained of. Remedies were every where proposed, but no serious attempt was made to apply them. France was, in these respects, in the same condition with other nations. A system, however, that had nothing to support it but the authority of the past time, and the inactivity of the present, was not likely to maintain itself long against the spirit of reform which became so general in that country at the commencement of the Revolution. This system, too, beside the other objections to it, had the misfortune to appear connected with all the abominations of the feudal times. The abolition of it, therefore, was resolved on ; and it would have been happy for France and for Europe, if every thing which was then destroyed had been replaced by as solid and useful a structure as that which we are going to describe. In the reformation propos- ed, two principal objects were kept in view. The first was the establishment of a natural standard for the measures of linear extension, and of course for the measures of all other quantities. The second was, to render the computation of those measures subject to the same arithmetical system that is used in other calculations. For this purpose, the unit of measure was to be divided decimally, and to be multiplied decimally, in order to constitute the other measures which it might be necessary to em- ploy. No fractions but decimal were to be used in expressing quantities of any sort ; and the great improvement of having but one arithmetical scale for reckoning integers and fractions of every kind, was in this way to be introduced ;—an improve- ment so obvious, and withal so little difficult, that it is matter of surprise that it should not have been attempted till near a thousand years after decimal arithmetic itself was first introduced into Europe.

In treating of this reform, however, we cannot

help remarking that the French academicians,
though freed at the moment we now speak of, like
the rest of their countrymen, from the dominion of
that *inertia* which reigns so powerfully both in the
natural and moral world, and gives the time that is
past such influence over that which is to come ;
though delivered from the action of this force, in a
degree that was perhaps never before exemplified,
they may be accused, at least in one instance, of
having innovated too little, and of having been too
cautious about departing from an established practice, though reason was by no means on its side.
What we allude to, is the system of arithmetical
computation, in which they resolved to adhere to
the decimal scale, instead of adopting the duodecimal, which, from the nature of number, is so evidently preferable. This preference, we believe, is
generally admitted in theory ; and there can be no
doubt, that a rational being, conversant with the
nature of number, if called on to choose his own
arithmetical system, and having no bias from custom, prejudice, or authority, would not hesitate a
moment about adopting the duodecimal system in
preference to the decimal, and, as we think, in
preference to all other systems whatsoever. The
property of the number twelve, which recommends
it so strongly for the purpose we are now considering, is its divisibility into so many more aliquot
parts than ten, or any other number that is not

much greater than itself. Twelve is divisible by 2, by 3, by 4, and by 6 ; and this circumstance fits it so well for the purposes of arithmetical computation, that it has been resorted to, in all times, as the most convenient number into which any unit either of weight or of measure could be divided.

The divisions of the *As*, the *Libra*, the *Jugerum*, the *Foot*, are all proofs of what is here asserted ; and this advantage, which was perceived in rude and early times, would have been found of great value in the most improved state of mathematical science. Ten has indeed no advantage as the radix of numerical computation, and has been raised to the dignity which it now holds, merely by the circumstance of its expressing the number of a man's fingers. They who regard science as the creature of pure reason, must feel somewhat indignant, that a consideration so foreign and mechanical should have determined the form and order of one of the most intellectual and abstract of all the sciences.

The duodecimal scale would no where have been found of greater use than when applied to the circle, the case in which the decimal division is liable to the strongest objections. The number by which the circumference of the circle is expressed, ought not only to be divisible into four integer parts, (as in the French system,) but also into six ; for the sixth part of the circumference, having its chord equal to the radius, naturally falls, in the construction of instruments, and in the computations of trigonometry, to be expressed by an integer number. According to the decimal division of the quadrant, the sixth part of the circumference not only is without an integer expression, but the decimal fraction by which it is measured is one that runs on continually without any termination. This is at least a deformity that arises from the rigid adherence to the decimal division; and it is probably the main cause why that division has been found so difficult to introduce into trigonometrical and astronomical calculation. In astronomical tables, we believe it has never been adopted. *

The adopting of twelve for the radix of the arithmetical scale would have obviated all these difficulties; it could have been extended with equal ease

---

* Supposing the decimal division to be extended to the circle, instead of dividing the quadrant into 100, and the circumference into 400, as the French have done, it would have been better, perhaps, to have divided the sixth part of the circumference into 100, the quadrant of course into 150, and the whole circumference into 600. This would have given an easy expression for the three great *natural* divisions of the circumference into 6, 4, and 2; and would have denoted the whole by a number (600) which does not violate the strict rule of dividing by the powers of 10, any more than 400 does. The advantages of the decimal and sexagesimal systems would by this means have been in a great measure united.

to quantities of every kind ; and the introduction of it would not have been accompanied with any present inconvenience of such magnitude as should have deterred geometers from making the attempt. We have lately seen a manuscript containing the system of duodecimal arithmetic pursued into all its detail.   Two new names are necessary for the numbers eleven and twelve ; and the whole arithmetical language, for the numbers above ten, is consequently changed, but in a manner so analogical, as to remove all difficulty, whether in the contrivance or in the acquisition of this new vocabulary. The arithmetical characters must also undergo an entire change ; the first eleven letters of the Greek alphabet are adopted in the scheme to which we refer ; and by means of them and the cypher, which is still retained, the notation proceeds by rules that are easy, and well known.

We regret, therefore, that the experiment of this new arithmetic was not attempted.   Another opportunity of trying it is not likely to occur soon. In the ordinary course of human affairs, such improvements are not thought of; and the moment may never again present itself, when the wisdom or delirium of a nation shall come up to the level of this species of reform.

But, to return to what respects the natural and universal standard of measure, we must remark, that the fixing on such a standard, and the abolition of the present diversity of weights and measures, was an object that very early drew the attention of the Constituent Assembly. It was proposed in that assembly by M. de Talleyrand, and decreed accordingly, that the king should be intreated to write to his Britannic Majesty, to engage the Parliament of England to concur with the National Assembly in fixing a natural unit of weights and measures; that, under the auspices of the two nations, an equal number of Commissioners from the Academy of Sciences and the Royal Society of London, might unite in order to determine the length of the pendulum in the latitude of 45°, or in any other latitude that might be thought preferable, and to deduce from thence an invariable standard of measures and of weights. This decree passed in August 1790. The Academy named a Commission, composed of Borda, Lagrange, Laplace, Monge, and Condorcet; and their report is printed in the Memoirs of the Academy for 1788. * Three different units fell under the consideration of these philosophers; to wit, the length of the pendulum, the quadrant of the meridian, and the quadrant of the equator. If the first of these was to be adopted, the commissioners were of opinion, that the pendulum vibrating seconds in the parallel of 45° deserved the preference, because

---

* Published 1791.

it is the arithmetical mean between the like pendulums in all other latitudes.   They observed, however, that the pendulum involves one element which is heterogeneous, to wit, *time*, and another, which is arbitrary, to wit, the division of the day into 86,400 seconds.   It seemed to them better that the unit of length should not depend on a quantity of a kind different from itself, nor on any thing that was arbitrarily assumed.

The commissioners, therefore, were brought to deliberate between the quadrant of the equator, and the quadrant of the meridian ; and they were determined to fix on the latter, because it is most accessible, and because it can be ascertained with most precision.   The quadrant of the meridian then was to be taken as the real unit ; and the ten-millionth part of it, being thought of a convenient length, was to be taken, in practice, for the unit of linear extension.   At the same time, the ordinary division of the circle into 360° was to be abandoned, and the decimal division introduced ; the fourth part of the circumference being divided, not into 90, but into 100 equal parts ; these parts into ten, and so on.   With regard to the above determination, we must be permitted to remark, that the reasons for rejecting the pendulum are by no means completely satisfactory.   The consideration, that time is a heterogeneous element, is too abstract and metaphysical to influence one's choice in a matter that is merely practical. The arbitrary element introduced by the division of the day into seconds, is perhaps an objection of more weight, were it not balanced by an equal objection in the case of the standard which has been actually adopted. That standard, in effect, is not the quadrant of the meridian, but the ten-millionth part of that quadrant; and ten million is without doubt a number just as arbitrary, and as far from being suggested by any natural appearance, as 86,400, the number of seconds into which the day is divided. It is impossible, indeed, whatever standard be adopted, to proceed without the use of some arbitrary division that must be determined by our conveniency, and not at all by the nature of the thing itself. Whether we take the quadrant of the meridian, or the radius of the globe, as Cassini long ago proposed, for the unit with which all measures are to be compared, the portion of that standard which we can convert into a rod of brass or platina, to be preserved in our museums, or to be employed in actual mensuration, must be a matter of arbitrary determination. The real unit or standard that is used in practice, must always involve in it a similar assumption; and its doing so can never afford a good reason for rejecting one standard, and preferring another.

It may be further alleged against the choice of the French Commissioners, that there is in

the unit which they have fixed on, something that
is even worse than an arbitrary element, one which
is hypothetical, and accompanied with some degree
of uncertainty. The quadrant of the meridian
itself is not the immediate object of mensuration, at
least in its whole length. That length is concluded from the mensuration of a part, on the supposition that the meridian is an ellipsis, and that the
ratio of its axes to one another is known. It is
supposed, too, that the meridians are similar and
equal curves; so that, in whatever place of the
world an arch of the meridian is measured, the
quadrant deduced from it will be of the same magnitude. It is well known that these suppositions
are not rigorously true, and, what is most material
of all, that a very large arch, or several different
arches of the meridian, must be measured, before
the length of the whole can be determined with
tolerable exactness. In all these respects, the pendulum, in the latitude of 45", seems to us to have
the decided preference above all others. The determination of it involves no theory, at least none
about the conclusions of which the slightest doubt
is entertained : it is at all times easily examined;
and nature constantly holds out the prototype with
which our standard may be compared, and from
which, if lost, the knowledge of it may easily be recovered.

For these reasons, notwithstanding our profound

respect for the genius and talents of the five academicians above named, we acknowledge that we are
unable to acquiesce in the arguments by which they
appear to have been determined.

But however this be, it cannot be questioned,
that after the French academicians had laid down
their plan, their method of carrying it into execution was most expeditious and accurate. They did
not wait for the determination of the English
government; and we are not informed what steps
were taken in consequence of the decree of the
Convention which has already been mentioned.
Indeed, though none of those events had taken
place which have since alienated the two nations so
entirely from one another, the slow proceedings of
an old government like ours, could never have kept
pace with the ardour of reform by which at that
moment the whole of the French nation seemed to
be animated.   The first step of the Commissioners
was to set about the measurement of the largest
arch of the meridian, which the extent of the dominions of France, or of its allies, rendered accessible to them.   This arch, extending from Dunkirk to Barcelona, contains something more than
nine degrees and a half, six of them to the north,
and three and a half to the south, of the mean
parallel of 45°.

The greater part of this arch had been already
measured more than once, and the length of it was

now to be determined for the third time. Picard had measured the degree between Paris and Amiens in the year 1670 ; and this arch was continued to Dunkirk on the north, and Collioure on the south, by Cassini and Lahire in the beginning of the last century, and in the end of that which preceded it. Doubts having arisen concerning the accuracy of this measurement, Cassini de Thury, and Lacaille, were charged with the verification of it in 1739. The measurement of these mathematicians, conducted with great skill and attention, was as correct as the construction of mathematical instruments, at the time when it was made, would permit ; and from it, no doubt, the length of the quadrant of the meridian might have been computed. The improvements, however, that had been made in the construction of instruments, gave reason to hope that a degree of accuracy considerably greater might now be attained.

Six different commissions for carrying all the parts of the plan into execution, that is, for ascertaining the unit of weight, the length of the pendulum, &c. &c. were appointed ; and the principal part, to wit, the measurement of the arch just mentioned, was committed to Mechain and Delambre, who began their operations in summer 1792. The instruments which these mathematicians were to employ, both in their astronomical and geodetical operations, were the repeating circles of Borda,

which had been so successfully used by the French
astronomers in their operations for connecting the
observatories of Greenwich and Paris in 1787.
Four new instruments of that kind, and of somewhat larger dimensions, were executed by Lenoir,
a very skilful artist, and put into the hands of
Mechain and Delambre ; the former of whom was
to measure the part of the arch between Barcelona
and Rhodez, about 170,000 toises in length, while
the latter measured the remaining arch of 380,000
toises, between Rhodez and Dunkirk.

The reason of this unequal partition of the labour
was, that the southern or Spanish part of the line
was entirely new, and therefore seemed likely to
present more difficulty than the northern part,
which had been already measured twice, and where
the time taken up in selecting stations, and laying
the general plan of the survey, was likely to be
spared.   It was soon found, however, that the difficulties which obstructed the work in the north,
arising from very different causes from any which
had been hitherto experienced, were much greater
than in the south ; and that, in those moments of
popular ferment and agitation, the neighbourhood
of the metropolis was the place where vexation,
and even danger, were most likely to be encounter-
ed.—The work, indeed was undertaken at a time
singularly unpropitious to the tranquil pursuits of
science.   The people, in the heat of the revolu-

tionary proceedings, jealous of whatever they did not understand, saw, in the astronomers and their apparatus, nothing but causes of alarm. When they observed men professing to be employed in a service which they could not comprehend, and accompanied with instruments of so mysterious a form, they thought the whole was a pretence under which the enemies of the people concealed their machinations. Delambre was more than once obliged to stop in the towns and villages, and to read a lecture to the incredulous multitude on the nature of his astronomical apparatus, and the purposes to which it was to be applied. The magistrates afforded him all the protection in their power ; but at that moment their power was of slow and precarious operation. The coolness and intrepidity of the French astronomer, added to unexampled patience, were the principal means of extricating him from his difficulties : but his danger was often imminent ; and he appears sometimes to have heard the dreadful words, which, as an eloquent author has expressed it, " were the last sounds that vibrated in the ear of many an unhappy victim."

Mechain was more fortunate. He was once stopped in the vicinity of Paris ; but having got to a distance from the capital, he met with no further disturbance. Both astronomers suffered much inconvenience in the prosecution of their work, from the depreciation of the assignats, in consequence of which they were often reduced to great necessity, and were deserted by most of their assistants. Delambre also met with persecutions from the tribunal, at that time so formidable to worth and innocence, wherever they were found. The following is an extract from the register of the Committee of Public Safety ; and in the censure which it pronounces on Delambre and his associates, posterity will never fail to recognise the most honourable testimony in their favour.

" The Committee of Public Safety, considering how much it imports to the amelioration of the public mind, that those who are charged with the government do not delegate any function, or give any commission, but to men worthy of confidence, from their republican virtues, and their hatred of Kings, after having consulted with the members of the Committee of Public Instruction, particularly occupied about the operation of weights and measures, decrees, that Borda, Lavoisier, Laplace, Coulomb, Brisson, and Delambre, shall cease from this day to be members of the Commission of Weights and Measures, and shall immediately lodge, in the hands of the remaining members of the said Commission, their instruments, calculations, notes, &c. together with an inventory."

To this extract are annexed the names of Barrere, Robespierre, Billaud-Varrennes, Couthon, Collot-d'Herbois, &c. To have fallen under the

displeasure of such men, will be allowed matter of
no ordinary praise, when it is considered how
deadly their hatred was. We are glad to think,
that at least four of those who merited this praise,
have lived to see the time when its true value could
be safely acknowledged. The preceding decree is
dated in the third year of the republic ; and, happily for France, the power of the Committee of
Public Safety did not extend far beyond that period.
The operations concerning the measurement, though
discontinued for some time, were afterwards resumed and completed, and the one end of the arch connected with the other by a series of triangles.

Two bases were also measured, one at Melun by
Delambre, of 6075.9 toises, another at Perpignan,
by Mechain, of 6006.248 toises. It appears from
Delambre's account, that when the second of these
bases was inferred from the first, it was found only
about ten or eleven inches shorter than it turned
out to be by actual measurement. When it is considered that the distance between them is about
360330 toises, or something more than 436 English miles, it will be admitted that this coincidence
is a proof of extreme accuracy. At the same time
we should have expected, that in the extent of so
long a line, one or two more bases should have been
measured. Colonel Mudge, in the conduct of his
survey, though the angles of the triangles are taken
with an exactness that cannot be exceeded, seems to

think, that the mensuration of a base for every hundred miles, if not absolutely necessary, is at least extremely desirable.

The observations, when finished, were laid before a Commission formed of members of the National Institute, and a great number of learned and scientific men, from Germany, Denmark, Holland, Italy, &c. who had accepted the invitation given them to assist in the solution of this great problem. The manner of proceeding before this commission was such, as to give the utmost degree of authenticity and correctness to all the parts of the work. The three angles of every triangle were separately examined ; and after comparing the different observations of each angle, with all the circumstances entered into the original note-books and registers, and attending to all the explanations furnished by the two observers themselves, the commissioners drew up the table of triangles, such as it is given at the end of this volume, and such as was to be used in all the subsequent calculations.   These calculations were all separately carried on by four different persons, Tralles, Van Swinden, Legendre, and Delambre himself.   Each gave in his own calculations ; and their differences, if there were any, being again examined, the result was finally agreed on.   The observations for the azimuth were subjected to the like examination ; and, from all these combined, the length of the arch of the meridian was inferred. The observation for the latitude made at Dunkirk, Paris, Evaux, Carcassonne and Montjouy, were also produced ; so that the celestial arch became known. The comparison of the two gave, for the compression of the earth $\frac{1}{334}$; for the quadrant of the meridian 5130740 toises ; and, consequently, for the metre 443.295986 lines.

During this interval, Mechain and Delambre had each fixed the latitude of his observatory by no less than 1800 observations, in order to determine from thence the latitude of the Pantheon, which was a little to the westward of the meridian, and the vertex of four of the triangles. These observations agreed with one another to the sixth part of a second.

The special commission for determining the length of the metre consisted, at this time, of Van Swinden, Tralles, Laplace, Legendre, Siscar, Mechain, and Delambre. Their report, drawn up by Van Swinden, is dated the sixth of Floreal, in the year VII, that is 1799. The present work, though begun in the year following, did not make its appearance till six years afterwards, having been interrupted or delayed by various circumstances, and particularly by Mechain's journey into Spain, for the purpose of continuing the meridian through the Balearic isles. After suffering many hardships and disappointments, this excellent astronomer had nearly brought the work to a conclusion, when an epidemical fever, added to the fatigue he had endured, carried him off at Castellan de la Plana in Valencia, in the year 1805. *

Such is the historical notice prefixed to this work; and the remainder of the preliminary discourse is chiefly taken up in explaining the formulas and tables employed in the reduction and correction of the observations. These formulas deserve to be studied with attention by all concerned in similar operations, to whom also we would take the liberty of recommending a smaller tract by Delambre, entitled, Méthodes Analytiques pour la détermination d'un Arc du Méridien, where the principles of these reductions are explained in a manner that renders it one of the most useful books on the

---

* This part of the work, we are enabled to state, though not on the authority of the volume before us, has not been abandoned. During the late negotiation at Paris, a passport was granted by the British government, at the request of the French minister, to M. Biot, charged with the operation of extending the meridian of Paris to the Balearic isles. The whole arch, if it be continued to Majorca, will then amount to about twelve degrees, bisected nearly by the parallel of 45. This is by much the longest arch yet measured on the surface of the earth: it is situated most favourably for the exact determination of the quadrantal arch ; and as part of it is across the sea, it may bring into view some important facts concerning the effect of difference of density in disturbing the direction of the plumb-line.

higher parts of practical geometry which has yet appeared.

We have already remarked, that the repeating circle of Borda was the instrument with which all the angles were observed in the course of this work. That instrument, though of no greater radius than seven inches, by the facility which it affords of measuring any proposed multiple of the angle observed, is capable of giving a mean result of much greater exactness than could be expected from its dimensions. To give an idea of the manner in which these results were obtained, we may take for an example, the angle between Watten and Cassel, observed at Dunkirk. The quadruple of that angle was found to be 187.147 gr., from which the angle itself was deduced, 46.78655 gr. The divisions, here, we must remark, are decimal, the quadrant being divided into 100, which are called *grades*. The same angle, multiplied ten times, was found = 467.868 gr., from which the angle itself appeared = 46.7868 gr. The multiplication was then carried to twelve, fourteen, &c. as far as twenty times the angle, which last was 935.731 gr., and the angle itself therefore = 46.78655 gr. It is very remarkable how near all these determinations come to one another; and the same is true of several more given by Delambre, but omitted here. From the mean of all these results, the true value

of the angle is finally determined.   It is the peculiar advantage of the instrument, that it allows these
repetitions, or multiplications, of an angle, to be
made easily, and in a very short space of time.   In
the circle which Delambre used, he tells us that he
could only read off from the *vernier* directly, as far
as hundredths of a decimal degree, the third decimal being set down by estimation.   The instrument, therefore, did not go to less than 32″ of the
common division, which is no very great degree of
exactness, and is much exceeded by many of the
sextants made in London for the ordinary purposes
of navigation.   Notwithstanding this, in consequence of the repeated multiplications of the angle,
we find a series of results obtained, that do not differ by more than a second from one another.

The repeating circle has, however, some disad-
vantages, at least when compared with certain other
instruments, to which it may be proper to advert.
Like our ordinary sextants, when an observation is
made with it, it is placed, not horizontally, but in
the plane of the three objects of which the bearings
are to be ascertained ; and, therefore, the altitude
of each of these objects, or its zenith distance, must
be observed, in order that the angle may be reduced to the plane of the horizon.   The determination of every horizontal angle, therefore, requires
that of three different angles ; which, no doubt,
must be accounted an imperfection in this method

of surveying, compared with that which places the
instrument immediately in the plane of the horizon,
and so gives the result directly, and without any reduction whatsoever.

Besides a reduction to the plane of the horizon,
another reduction is necessary to bring the observed angles to the true angles, at the centres of the
signals. For this reduction, as well as for the preceding, Delambre has given rules and formulæ, by
which they may be calculated with great accuracy,
and with all the expedition that the nature of the
thing will admit.—It were, nevertheless, very desirable, that these reductions, as well as the former,
should be avoided, by placing the instrument with
which the angles are taken, exactly at the angular
point.—This method has been generally followed
in the trigonometrical survey of England, where no
pains have been spared to place the theodolite in
the same spot that was occupied by the centre of
the signal : and from thence results great additional accuracy, as well as additional simplicity, in
conducting the calculations.   It is the same with
respect to the reduction to the horizontal plane.
The great theodolite, first employed by General
Roy, and now in the hands of Colonel Mudge, is
always placed horizontally ; and hence a great deal
of labour in calculation is saved, and many sources
of inaccuracy are entirely avoided.   In no other
survey, we believe, has the work in the field been

conducted so much with a view to save that in the
closet, and at the same time to avoid all those
causes of error, however minute, that are not essentially involved in the nature of the problem.   The
French mathematicians trust to the correction of
those errors ; the English endeavour to cut them
off entirely ; and it can hardly be doubted that the
latter, though perhaps the slowest and most expensive, is by far the safest proceeding.

The principal defect of the reflecting circle, we
believe to consist in the small power of the telescope
which it bears ; an imperfection inseparable from
an instrument of such a size as can be held in the
hand.   The accuracy of the observations is necessarily limited, by the imperfections of the telescope,
to whatever degree of minuteness the divisions on
the limb of the instrument may be carried.   If an
object that subtends an angle of 3 or 4 seconds is
the least that is distinctly visible through the telescope, an angle can never be measured nearer than
3 or 4 seconds, even if you can read off to a single
second.   The want of sufficient light, also, in the
field of vision, seems to have occasioned considerable
difficulty, and may have produced some inaccuracy
in the observations.   The signals, when the distance was considerable, were not always distinctly
seen ; and the construction of them became, on
that account, an object of very great attention.
Delambre made his signals in the form of a pyra-

mid, truncated at the top, of the height of nearly $\frac{3}{20000}$ of the distance ; so that they subtended an angle of about 30″. They were built of wood, and the base was a third part of the height. The powerful telescopes used by General Roy and Colonel Mudge in the trigonometrical survey of England, relieved them from all difficulties of this kind ; as a simple mast, or staff, as it is called, with the ordinary illumination of a clear day, can be seen distinctly through the telescope of the great theodolite at the distance of 15 or 16 miles. The advantage of these large glasses was experienced by the French Academicians, when they met with the committee of the Royal Society of London, in order to unite the surveys made in France and in England, for the purpose of ascertaining the relative position of the observatories of Greenwich and Paris, as we had occasion to remark in our review of the trigonometrical survey.

It probably arose from the same cause, in some measure, that the signals made by lamps in the night were not found to answer with the French astronomers. The difficulty of illuminating the hairs in the focus of the telescope, is the impediment chiefly complained of. Such signals, however, were sometimes employed ; and Delambre mentions a curious phenomenon which he observed, more than once, to accompany them. This was a kind of undulation, which made the apparent place

I

of the light oscillate very sensibly about the true, as
was particularly remarked at Mont-Martre, on the
15th of August 1792.   Something similar to this,
he says, he had met with in the case of the ordinary signals viewed during the day.

" These oscillations," says he, " I have sometimes thought would remain suspended for a few
minutes, when the apparent object was at its greatest distance from the real.   I will not, however,
answer for this fact, which I have not been able to
investigate sufficiently.   I have also been disposed
to think that there existed a lateral refraction.
The only way to guard against the effects of such
illusions, is to wait the arrival of fine weather, and
to repeat the observations in circumstances as unlike
as possible."

The confidence which the French astronomers
place in the repeating circle, is such, that they have
not, in the course of this work, had recourse to a
zenith sector, or any of the more delicate instruments of astronomy, for the purpose of determining
the differences of latitude, or the amplitudes of the
celestial arches, corresponding to the lines measured
on the surface of the earth.   This, we confess, appears to us not a little extraordinary, though we
must, at the same time, remark, that this reliance
on the repeating circle is confirmed by the opinion
of the Swedish astronomers, who have lately mea-

sured the degree in Lapland anew. They used no other instrument but the repeating circle ; and Lalande says, in his sketch of the history of astronomy for 1805, that they thought that instrument less liable to error than a zenith sector of nine feet radius, such as was used by Maupertuis and his colleagues.

On the whole, when we compare the methods and instruments used in the French and in the British survey, though we see many circumstances that would induce us to give a preference to the latter ; yet, when we consider the results of each, they seem, in exactness and consistency, to approach very near to an equality. We are not sure to what this should be ascribed ; nor can we form a decided opinion till the reductions of the distances to the meridian are given. It may be, that the great expedition with which the repeating circle is used, and the vast number of observations which it admits of being made in a short space of time, may balance the greater size and more exquisite division of the theodolite and the sector employed by our observers ; instruments which, in themselves, are perhaps the most perfect that have ever been constructed. Indeed, the expedition with which the repeating circle can be used, is one of its greatest advantages. It is such, that, in the space of five years, two observers, with very few as-

10

sistants, * and in the midst of innumerable interruptions and vexations, completed a survey of 181 triangles, extending along a line more than 600 miles in length, and this, together with the measurement of two bases, and a vast number of observations for determining azimuths, and the latitudes of five different stations.

The volume before us is far from completing the account of these operations, or at least of the conclusions deduced from them. The part we have already considered, which forms the preliminary discourse, is followed by 510 pages, in which a detailed account is given of the observations made at each station, and of all the circumstances by which their accuracy can be affected. The triangles included in these observations consist of 115 principal, and 66 subsidiary triangles. At the end of the detailed account just mentioned, these triangles are reduced into a table, that exhibits, at one view, every thing concerning their angles,.and the necessary reductions. The first column of this table

---

* Delambre was assisted by *Bellet,* an astronomical instrument maker of great zeal and intelligence, who adhered to him in all his difficulties, and remained, when the smallness of the allowance from government deprived him of all his other assistants. Francis Lalande also shared with him in the labour of the survey for a considerable time. In the southern survey, Mechain was assisted by M. Tranchot, an engineer of great ability and experience.

contains the angles of every triangle as observed
and reduced to the horizon ; the numbers here
given being the means that were struck by the
commissioners in the manner already described.
The second column contains the spherical excess
for each angle, by comparing which with the sum
of the three angles of the triangle, we have a measure of the error committed in the measurement
of the three angles, which rarely amounts to $1''$ or
$1\frac{1}{2}''$. In the third column are given the true spherical angles corrected for the error of observation,
according to a principle previously explained. In
the fourth column, these are reduced to the rectilineal angles contained by the chords of the arches,
or of the sides of the spherical triangles. The last
column of all contains the mean angles, as they are
here called, that is, the observed angles first corrected for the error of observation, and afterwards diminished each by one third of the spherical excess,
so that those of each triangle amount precisely to
180 degrees, and are thus prepared for computation,
according to the theorem of Legendre that was
mentioned in our account of the trigonometrical
survey of England. Delambre has gone through
the great labour of calculating the sides of these
triangles, and also their reduction to the meridian,
by the three different methods that correspond to
the nature of the three last columns. These results, however, are not given in the volume before

us. They are reserved for that which is to follow ; and we have no doubt, when they shall appear, will give us new reasons to applaud the skill, the accuracy, and the patience of Delambre and his associates.

Though the formation of the above table, and the arrangement of the whole volume, are the work of Delambre, yet a large part of it, containing the observations of Mechain, is given in the words of that astronomer.

The present volume does not enable us to state any thing with respect to the length of the segments into which the arch of the meridian was divided. Some of these results, however, we have learnt from the memoirs that have been published on the same subject in the volumes of the National Institute. They appear to be curious ; and we take notice of them here, as indicating irregularities very similar to those that Colonel Mudge met with in the meridional arch which he measured between the Isle of Wight and Clifton in Yorkshire.

In the course of their survey, the French astronomers determined the latitudes of five different points of the meridional arch with great exactness, viz. Dunkirk, 51° $2'$ $10''$; Paris, at the Pantheon, 48° $50'$ $49\frac{3}{4}''$; Steeple of Evaux, 46° $10'$ $42''$; Tower of St Vincent at Carcassonne, 43° $12'$ $54''$; Tower of Montjouy at Barcelona, 41° $21'$ $45''$. The amplitudes of the arches thus found, being

compared with the terrestrial measurements, led to
some results that were unexpected, and that are
certainly highly deserving of attention. It appears that the length of the degree of the meridian,
though it decrease constantly on going from the
north to the south, as it ought to do, does in fact
decrease very irregularly. Toward the northern
extremity of the arch the decrease is slow, and at
the rate of not more than four toises in the degrees
that lie between Dunkirk and Evaux. From
Evaux to Carcassonne the degrees diminish rapidly, at the rate of 30 or 31 toises; and from Carcassonne to Barcelona the diminution becomes
again much slower, and is about 14 toises to a degree.

This irregularity in the differences of the degrees does not arise from a cause that is apparent
on the surface. It very much resembles that
which was experienced by Colonel Mudge as he
went northward from the coast of the channel,
when he found that the degrees, instead of increasing, came to diminish about the middle of
the arch. In both cases, we may suspect the effect to have arisen, partly from the vicinity of the
sea, partly, perhaps, from inequalities of density
under the surface, or other irregularities in the su-
perficial part of the globe. From whatever causes
they arise, the repetition of operations, such as those
we are now treating of, is what alone can be ex-

pected to throw new light upon the subject.   Additional experiments on the attraction of mountains would probably tend to the same object, and might lead to other valuable conclusions.

We cannot finish our account of these scientific operations, without expressing our wishes that the uniformity of measures and of weights were introduced into our own, and into every other civilized country.   The difficulty is not so great as we are apt to think, when we consider the matter at a distance ; and, to effect it, requires, in reality, nothing but for the legislature to say, it shall be done. As to the standard to be adopted, though we think the pendulum would have afforded the most convenient, yet, when one has been actually fixed on and determined, that circumstance must greatly outweigh every other consideration.   The system adopted by the French, if not absolutely the best, is so very near it, that the difference is of no account.   In one point it is very unexceptionable ; it involves nothing that savours of the peculiarities of any country ; insomuch, as the Commissioners observe, that if all the history which we have been considering were forgotten, and the results of the operations only preserved, it would be impossible to tell with what nation this system had originated. The wisest measure, therefore, for the other nations of Europe, is certainly to adopt the metrical system of the French, with the exception, perhaps, of the division of the circle, where the number 600, as mentioned above, might be conveniently substituted for 400. It would not be necessary to adopt their names, which might not assort very well with the sounds that compose the languages of other nations. But the *mètre*, by whatever name it may be called, ought to be adopted as the unit of length, and all the other measures of linear extension derived from it by decimal multiplication and division. It is true that this cannot be done, especially in our own case, without a certain sacrifice of national vanity ; and the times do not give much encouragement to hope that such a sacrifice will be made. The calamities which the power and ambition of the French government have brought on Europe, induce us to look with jealousy and suspicion on their most innocent and laudable exertions. We ought not, however, to yield to such prejudices, where good sense and argument are so obviously against them. In a matter that concerns the arts and sciences only, the maxim may be safely admitted, *Fas est et ab hoste doceri.*

FINIS.

REVIEW

OF

LAPLACE, TRAITÉ DE MECANIQUE

CELESTE.

REVIEW

OF

LAPLACE, MECANIQUE CELESTE. *

ASTRONOMY is distinguished by several great and
striking characters, which place it decidedly at the
head of the physical sciences.   The objects which
it treats of cannot fail to impart to it a degree of
their own magnificence and splendour ; while their
distance, their magnitude, the steadiness and regularity of their movements, deeply impress the imagination, and afford a noble exercise to the understanding.   Add to this, that the history of astronomy is that which is best marked out in the progress of human knowledge.   Through the darkness
of the early ages, we perceive the truths of this
science shining as it were by their own light, and
scattering some rays around them, that serve to
discover a few definite objects amid the confusion
of ancient tradition,—a few fixed points amid the
uncertainty of Greek, Egyptian, or even Hindoo

---

* From the Edinburgh Review, Vol. XI. (1808.)—ED.

mythology.   But what distinguishes astronomy the
most, is the perfect explanation which it gives of
the celestial phenomena.   This explanation is so
complete, that there is not any fact concerning the
motions of the heavenly bodies, from the greatest
to the least, which is not reducible to one single
law—the mutual gravitation of all bodies to one
another, with forces that are directly as the masses
of the bodies, and inversely as the squares of their
distances.   On this principle Sir Isaac Newton
long ago accounted for all the great motions in our
system ; and, on the same principle, his successors,
after near a century of the most ingenious and elaborate investigation, have explained all the rest.
The work before us brings those explanations into
one view, and deduces them from the first principles of mechanics.   It is not willingly that we have
suffered so much time to elapse without laying before our readers an analysis of a work the most important, without doubt, that has distinguished the
conclusion of the last or the commencement of the
present century.   But the book is still, in some
respects, incomplete ; and a historical volume is yet
wanting, which, had we been in possession of it,
would have very much facilitated the task that we
have now undertaken to perform.   We know not
whether this volume is actually published.   In the
present state of Europe it may be a long time before
it can find its way to this country ; and, in the

mean time, our duty seems to require, that an account of the four volumes, which we possess, should no longer be withheld from the public.

Though the integral calculus, as it was left by the first inventors and their contemporaries, was a very powerful instrument of investigation, it required many improvements to fit it for extending the philosophy of Newton to its utmost limits. A brief enumeration of the principal improvements which it has actually received in the last seventy or eighty years, will very much assist us in appreciating the merit of the work which is now before us.

1. Descartes is celebrated for having applied algebra to geometry ; and Euler hardly deserves less credit for having applied the same science to trigonometry. Though we ascribe the invention of this calculus to Euler, we are aware that the first attempt toward it was made by a mathematician of far inferior note, Christian Mayer, who, in the Petersburgh Commentaries for 1727, published a paper on analytical trigonometry. In that memoir, the geometrical theorems, which serve as the basis of this new species of arithmetic, are pointed out ; but the extension of the method, the introduction of a convenient notation, and of a peculiar algorithm, are the work of Euler. By means of these, the sines and cosines of arches are multiplied into one another, and raised to any power, with a simplicity unknown in any other part of algebra, being

expressed by the sines and cosines of multiple arches of one dimension only, or of no higher power than the first. It is incredible of how great advantage this method has proved in all the parts of the higher geometry, but more especially in the researches of physical astronomy. As what we observe in the heavens is nothing but angular position, so if we would compare the result of our reasonings concerning the action of the heavenly bodies, with observations made on the surface of the earth, we must express those results in terms of the angles observed, or the quantities dependent on them, such as sines, tangents, &c. It is evident that a calculus which teaches how this is to be accomplished, must be of the greatest value to the astronomer. Besides, the facility which this calculus gives to all the reasonings and computations into which it is introduced, from the elementary problems of geometry to the finding of fluents and the summing of series, makes it one of the most valuable resources in mathematical science. It is a method continually employed in the Mécanique Céleste.

2. An improvement in the integral calculus, made by D'Alembert, has doubled its power, and added to it a territory not inferior in extent to all that it before possessed. This is the method of *partial differences*, or, as we must call it, of *partial fluxions*. It was discovered by the geometer just named, when he was inquiring into the nature of

the figures successively assumed by a musical string during the time of its vibrations.   When a variable quantity is a function of other two variable quantities, as the ordinates belonging to the different abscissæ in these curves must necessarily be, (for they are functions both of the abscissæ and of the time counted from the beginning of the vibrations,) it becomes convenient to consider how that quantity varies, while each of the other two varies singly, the remaining one being supposed constant.   Without this simplification, it would, in most cases, be quite impossible to subject such complicated functions to any rules of reasoning whatsoever.   The calculus of partial differences, therefore, is of great utility in all the more complicated problems both of pure and mixed mathematics : every thing relating to the motion of fluids that is not purely elementary, falls within its range ; and in all the more difficult researches of physical astronomy, it has been introduced with great advantage.   The first idea of this new method, and the first application of it, are due to D'Alembert : it is from Euler, however, that we derive the form and notation that have been generally adopted.

3. Another great addition made to the integral calculus, is the invention of Lagrange, and is known by the name of the *Calculus variationum*.   The ordinary problems of determining the greatest and least states of a given function of one or more va-

riable quantities, is easily reduced to the direct method of fluxions, or the differential calculus, and
was indeed one of the first classes of questions to
which those methods were applied. But when the
function that is to be a *maximum* or a *minimum*, is
not given in its form ; or when the curve, expressing that function, is not known by any other property, but that, in certain circumstances, it is to be
the greatest or least possible, the solution is infinitely
more difficult, and science seems to have no hold
of the question by which to reduce it to a mathematical investigation. The problem of the line of
swiftest descent is of this nature ; and though from
some facilities which this and other particular instances afforded, they were resolved, by the ingenuity of mathematicians, before any method generally applicable to them was known, yet such a
method could not but be regarded as a great *desideratum* in mathematical science. The genius of
Euler had gone far to supply it, when Lagrange,
taking a view entirely different, fell upon a method
extremely convenient, and, considering the difficulty of the problem, the most simple that could be
expected. The supposition it proceeds on is greatly more general than that of the fluxionary or differential calculus. In this last, the fluxions or
changes of the variable quantities are restricted by
certain laws. The fluxion of the ordinate, for example, has a relation to the fluxion of the abscissa

that is determined by the nature of the curve to which they both belong. But in the method of *variations,* the change of the ordinate may be any whatever ; it may no longer be bounded by the original curve, but it may pass into another, having to the former no determinate relation. This is the calculus of Lagrange ; and, though it was invented expressly with a view to the problems just mentioned, it has been found of great use in many physical questions with which those problems are not immediately connected.

4. Among the improvements of the higher geometry, besides those which, like the preceding, consisted of methods entirely new, the extension of the more ordinary methods to the integration of a vast number of formulas, the investigation of many new theorems concerning quadratures, and concerning the solution of fluxionary equations of all orders, had completely changed the appearance of the calculus ; so that Newton or Leibnitz, had they returned to the world any time since the middle of the last century, would have been unable, without great study, to follow the discoveries which their disciples had made, by proceeding in the line which they themselves had pointed out. In this work, though a great number of ingenious men have been concerned, yet more is due to Euler than to any other individual. With indefatigable industry, and the resources of a most inventive mind, he devoted

a long life entirely to the pursuits of science. Besides producing many works on all the different branches of the higher mathematics, he continued, for more than fifty years during his life, and for no less than twenty after his death, to enrich the memoirs of Berlin, or of Petersburgh, with papers that bear, in every page, the marks of originality and invention. Such, indeed, has been the industry of this incomparable man, that his works, were they collected into one, notwithstanding that they are full of novelty, and are written in the most concise language by which human thought can be expressed, might vie in magnitude with the most trite and verbose compilations.

5. The additions we have enumerated were made to the pure mathematics ; that which we are going to mention belongs to the mixed. It is the mechanical principle, discovered by D'Alembert, which reduces every question concerning the *motion* of bodies, to a case of equilibrium. It consists in this—If the motions, which the particles of a moving body, or a system of moving bodies, have at any instant, be resolved each into two, one of which is the motion which the particle had in the preceding instant, then the sum of all these third motions must be such, that they are in equilibrium with one another. Though this principle is, in fact, nothing else than the equality of action and reaction, properly explained, and traced into the

secret process which takes place on the communication of motion, it has operated on science like one
entirely new, and deserves to be considered as an
important discovery. The consequence of it has
been, that, as the theory of equilibrium is perfectly
understood, all problems whatever, concerning the
motion of bodies, can be so far subjected to mathematical computation, that they can be expressed in
fluxionary or differential equations, and the solution of them reduced to the integration of those
equations. The full value of the proposition, however, was not understood, till Lagrange published
his *Mécanique Analytique* : the principle is there
reduced to still greater simplicity ; and the connection between the pure and the mixed mathematics, in this quarter, may be considered as complete.

Furnished with a part, or with the whole of
these resources, according to the period at which
they arose, the mathematicians who followed Newton in the career of physical astronomy, were enabled to add much to his discoveries, and at last to
complete the work which he so happily began.
Out of the number who embarked in this undertaking, and to whom science has many great obligations, five may be regarded as the leaders, and
as distinguished above the rest, by the greatness of
their achievements. These are, Clairaut, Euler,
D'Alembert, Lagrange, and Laplace himself, the

author of the work now under consideration.   By
their efforts it was found, that, at the close of the
last century, there did not remain a single phenomenon in the celestial motions, that was not explained on the principle of Gravitation ; nor any
greater difference between the conclusions of theory, and the observations of astronomy, than the errors
unavoidable in the latter were sufficient to account
for.   The time seemed now to be come for reducing
the whole theory of astronomy into one work, that
should embrace the entire compass of that science
and its discoveries for the last hundred years : Laplace was the man, in all Europe, whom the voice
of the scientific world would have selected for so
great an undertaking.

   The nature of the work required that it should
contain an entire System of Physical Astronomy,
from the first elements to the most remote conclusions of the science.   The author has been careful
to preserve the same method of investigation
throughout ; so that even where he has to deduce
results already known, there is a unity of character
and method that presents them under a new aspect.

   The reasoning employed is everywhere algebraical ; and the various parts of the higher mathematics, the integral calculus, the method of partial
differences and of variations, are from the first outset introduced, whenever they can enable the author to abbreviate or to generalize his investiga-

11

tions. No diagrams or geometrical figures are employed ; and the reader must converse with the objects presented to him by the language of arbitrary symbols alone. Whether the rejection of figures be in all respects an improvement, and whether it may not be in some degree hurtful to the powers of the imagination, we will not take upon us to decide. It is certain, however, that the perfection of algebra tends to the banishment of diagrams, and of all reference to them. Lagrange, in his treatise of Analytical Mechanics, has no reference to figures, notwithstanding the great number of mechanical problems which he resolves. The resolution of all the forces that act on any point, into three forces, in the direction of three axes at right angles to one another, enables one to express their relations very distinctly, without representing them by a figure, or expressing them by any other than algebraic symbols. This method is accordingly followed in the Mécanique Céleste. Something of the same kind, indeed, seems applicable to almost any part of the mathematics ; and a very distinct treatise on the conic sections, we doubt not, might be written, where there would not be a single diagram introduced, and where all the properties of the ellipse, the parabola, and the hyperbola, would be expressed either by words or by algebraic characters. Whether the imagination would lose or gain by this exercise, we shall not at present stop to inquire. It is curious, however, to observe, that algebra, which was first introduced for the mere purpose of assisting geometry, and supplying its defects, has ended, as many auxiliaries have done, with discarding that science (or at least its peculiar methods) almost entirely. We say, almost entirely; because there are, doubtless, a great number of the elementary propositions of geometry, that never can have any but a geometrical, and some of them a synthetical demonstration.

The work of Laplace is divided into two parts, and each of these into five books. The first part lays down the general principles applicable to the whole inquiry, and afterwards deduces from them the motions of the primary planets, as produced by their gravitation to the sun. The second part treats, first of the disturbances of the primary planets, and next of those of the secondary.

In the first book, the theory of motion is explained in a manner very unlike what we meet with in ordinary treatises,—with extreme generality, and with the assistance of the more difficult parts of the mathematics,—but in a way extremely luminous, concise, and readily applicable to the most extensive and arduous researches. This part must be highly gratifying to those who have a pleasure in contemplating the different ways in which the same truths may be established, and in pursuing whatever tends to simplicity and generalization. The greater part of the propositions here deduced are already known ; but it is good to have them presented in a new order, and investigated by the same methods that are pursued through the whole of this work, from the most elementary truths to the most remote conclusions.

For the purpose of instructing one in what may be called the Philosophy of Mechanics, that is, in the leading truths in the science of motion, and at the same time, in the way by which those truths are applied to particular investigations, we do not believe any work is better adapted than the first book of the *Mécanique Céleste*, provided it had a little more expansion given it in particular places, and a little more illustration employed for the sake of those who are not perfectly skilled in the use of the instrument which Laplace himself employs with so much dexterity and ease.

From the differential equations that express the motion of any number of bodies subjected to the mutual attraction of one another, deduced in the second chapter, Laplace proceeds to the integration of these equations by approximation, in the third and the following chapters. The first step in this process gives the integral complete in the case of two bodies, and shows that the curve described by each of them is a conic section. The whole theory of the elliptic motion follows, in which the solution of Kepler's problem, or the expression of the true

anomaly, and of the *radius vector* of a planet, in
terms of the mean anomaly, or of the time, are particularly deserving of attention, as well as the difference between the motion in a parabolic orbit,
and in an elliptic orbit of great eccentricity.

In the greater part of this investigation, the
theorems are such as have been long since deduced
by more ordinary methods ; the deduction of them
here was however essential, in order to preserve
the unity of the work, and to show that the simpler
truths, as well as the more difficult, make parts of
the same system, and emanate from the same principle.   These more elementary investigations have
this further advantage, that the knowledge of the
calculus, and of the methods peculiar to this work,
is thus gradually acquired, by beginning from the
more simple cases ; and we are prepared, by that
means, for the more difficult problems that are to
follow.

The general methods of integrating the differential equations above mentioned, are laid down in the
fifth chapter, which deserves to be studied with particular attention, whether we would improve in the
knowledge of the pure or the mixed mathematics.
The calculus of variations is introduced with great
effect in the last article of this chapter.

A very curious subject of investigation, and one
that we believe to be altogether new, follows in the
next chapter.   In the general movement of a sys-

tem of bodies, such as is here supposed, and such,
too, as is actually exemplified in nature, every thing
is in motion; not only every body, but the plane of
every orbit. The mutual action of the planets
changes the positions of the planes in which they
revolve ; and they are perpetually made to depart,
by a small quantity, on one side or another, each
from that plane in which it would go on continually,
if their mutual action were to cease. The calculus
makes it appear, that the inclinations of these orbits in the planetary system is stable, or that the
planes of the orbits oscillate a little, backwards and
forwards, on each side of a fixed and immoveable
plane. This plane is shown to be one, on which, if
every one of the bodies of the system be projected
by a perpendicular let fall from it, and if the mass
of each body be multiplied into the area described
in a given time by its projection on the said plane,
the sum of all these products shall be a *maximum*.
From this condition, the method of determining
the immoveable plane is deduced ; and in the progress of science, when observations made at a great
distance of time shall be compared together, the reference of them to an immoveable plane must become a matter of great importance to astronomers.

As the great problem resolved in this first book
is that which is called the problem of the Three
Bodies, it may be proper to give some account of
the steps by which mathematicians have been gra-

276 REVIEW OF LAPLACE,

dually conducted to a solution of it so perfect as that which is given by Laplace. The problem is, —Having given the masses of three bodies projected from three points given in position with velocities given in their quantity and direction, and supposing the bodies to gravitate to one another with forces that are as their masses directly and the squares of their distances inversely, to find the lines described by these bodies, and their position, at any given instant.

The problem may be rendered still more general, by supposing the number of bodies to be greater than three.

To resolve the problem in the general form contained in either of these enunciations, very far exceeds the powers even of the most improved analysis. In the cases, however, where it applies to the heavens, that is, when one of the bodies is very great and powerful in respect of the other two, a solution by approximation, and having any required degree of accuracy, may be obtained.

When the number of bodies is only two, the problem admits of a complete solution. Newton had accordingly resolved the problem of two bodies gravitating to one another, in the most perfect manner; and had shown, that when their mutual gravitation is as their masses divided by the squares of their distances, the orbits they describe are conic sections. The application of this theorem and

its corollaries to the motions of the planets round
the sun, furnished the most beautiful explanation
of natural phenomena that had yet been exhibited
to the world ; and however excellent, or in some
respects superior, the analytical methods may be
that have since been applied to this problem, we
hope that the original demonstrations will never be
overlooked.   When Newton, however, endeavoured to apply the same methods to the case of a
planet disturbed in its motion round its primary
by the action of a third body, the difficulties were
too great to be completely overcome.   The efforts,
nevertheless, which he made with instruments,
that, though powerful, were still inadequate to the
work in which they were employed, displayed, in a
striking manner, the resources of his genius, and
conducted him to many valuable discoveries.   Five
of the most considerable of the inequalities in the
moon's motion were explained in a satisfactory
manner, and referred to the sun's action ; but beyond this, though there is some reason to think
that Newton attempted to proceed, he has not made
us acquainted with the route which he pursued.   It
was evident, however, that besides these five inequalities, there were many more, of less magnitude indeed, but of an amount that was often considerable,
though the laws which they were subject to were
unknown, and were never likely to be discovered
by observation alone.

It is the glory of the Newtonian philosophy, not to have been limited to the precise point of perfection to which it was carried by its author; nor, like all the systems which the world had yet seen, from the age of Aristotle to that of Descartes, either to continue stationary, or to decline gradually from the moment of its publication. Three geometers, who had studied in the schools of Newton and of Leibnitz, and had greatly improved the methods of their masters, ventured, nearly about the same time, each unknown to the other two, to propose to himself the problem which has since been so well known under the name of the Problem of Three Bodies. Clairaut, D'Alembert, and Euler, are the three illustrious men, who, as by a common impulse, undertook this investigation in the year 1747; the priority, if any could be claimed, being on the side of Clairaut. The object of those geometers was not merely to explain the lunar inequalities that had been observed; they aimed at something higher; viz. from theory to investigate all the inequalities that could arise as the effects of gravitation, and so to give an accuracy to the tables of the moon, that they could not derive from observation alone. Thus, after having ascended with Newton from phenomena to the principle of gravitation, they were to descend from that principle to the discovery of new facts; and thus, by the twofold method of analysis and composition,

to apply to their theory the severest test, the only
infallible criterion that at all times distinguishes
truth from falsehood.   Clairaut was the first who
deduced, from his solution of the problem, a complete set of lunar tables, of an accuracy far superi-
or to any thing that had yet appeared, and which,
when compared with observation, gave the moon's
place, in all situations, very near the truth.

Their accuracy, however, was exceeded, or at
least supposed to be exceeded, by another set produced by Tobias Mayer of Gottingen, and grounded
on Euler's solution, compared very diligently with
observation.   The expression of the lunar irregularities, as deduced from theory, is represented by
the terms of a series, in each of which there are
two parts carefully to be distinguished ; one, which
is the sine or cosine of a variable angle determined
at every instant by the time counted from a certain
epocha ; another, which is a coefficient or multiplier, in itself constant, and remaining always the
same.   The determination of this constant part
may be derived from two different sources ; either
from our knowledge of the masses of the sun and
moon, and their mean distances from the earth ; or
from a comparison of the series above mentioned,
with the observed places of the moon, whence the
values of the coefficients are found, which make
the series agree most accurately with observation.
Mayer, who was himself a very skilful astronomer,

had been very careful in making these comparisons, and thence arose the greater accuracy of his tables. The problem of finding the longitude at sea, which was now understood to depend so much on the exactness with which the moon's place could be computed, gave vast additional value to these researches, and established a very close connection between the conclusions of theory, and one of the most important of the arts. Mayer's tables were rewarded by the Board of Longitude in England ; and Euler's, at the suggestion of Turgot, by the Board of Longitude in France.

It may be remarked here, as a curious fact in the history of science, that the accurate solution of the problem of the Three Bodies, which has in the end established the system of gravitation on so solid a basis, seemed, on its first appearance, to threaten the total overthrow of that system. Clairaut found, on determining, from his solution, the motion of the longer axis of the moon's orbit, that it came out only the half of what it was known to be from astronomical observation. In consequence of this, he was persuaded, that the force with which the earth attracts the moon, does not decrease exactly as the squares of the distances increase, but that a part of it only follows that law, while another follows the inverse of the biquadrate or fourth power of the distances. The existence of such a law of attraction was violently opposed by Buffon, who

objected to it the want of simplicity, and argued
that there was no sufficient reason for determining
what part of the attraction should be subject to the
one of these laws, and what part to the other.
Clairaut, and the other two mathematicians, (who
had come to the same result,) were not much influenced by this metaphysical argument ; and the
former proceeded to inquire what the proportion
was between the two parts of the attraction that
followed laws so different.

He was thus forced to carry his approximation
farther than he had done, and to include some
quantities that had before been rejected as too
small to affect the result. When he had done this,
he found the numerator of the fraction that denoted the part of gravity which followed the new law,
equal to nothing ; or, in other words, that there was
no such part. The candour of Clairaut did not
suffer him to delay, a moment, the acknowledgment of this result; and also, that when his calculus was rectified, and the approximation carried
to the full length, the motion of the moon's apsides as deduced from theory, coincided exactly
with observation.

Thus, the lunar theory was brought to a very
high degree of perfection ; and the tables constructed by means of it, were found to give the
moon's place true to 30''. Still, however, there
was one inequality in the motion's motion, for

which the principle of gravitation afforded no account whatever. This was what is known by the name of the moon's acceleration. Dr Halley had observed, on comparing the ancient with modern observations, that the moon's motion round the earth appeared to be now performed in a shorter time than formerly; and this inequality appeared to have been regularly, though slowly, increasing; so that, on computing backward from the present time, it was necessary to suppose the moon to be uniformly retarded, (as in the case of a body ascending against gravity,) the effect of this retardation increasing as the squares of the time. All astronomers admitted the existence of this inequality in the moon's motion; but no one saw any means of reconciling it with the principle of gravitation. All the irregularities of the moon arising from that cause had been found to be periodical; they were expressed in terms of the sines and cosines of arches; and though those arches depend on the time, and might increase with it continually, their sines and cosines had limits which they never could exceed, and from which they returned perpetually in the same order. Here, therefore, was one of the greatest anomalies yet discovered in the heavens—an inequality that increased continually, and altered the mean rate of the moon's motion. Various attempts were made to explain this phenomenon, and those too attended with much intricate and laborious investigation.

To some it appeared, that this perpetual decrease in the time of the moon's revolution, must arise from that resistance of the medium in which she moves, which, by lessening her absolute velocity, would give gravity more power over her ; so that she would come nearer to the earth, would revolve in less time, and therefore with a greater angular velocity. This hypothesis, though so unlike what we are led to believe from all other appearances, must have been admitted, if, upon applying mathematical reasoning, it had been found to afford a good explanation of the appearances. It was found, however, on trial, that it did not ; and that the moon's acceleration could not be explained by the supposed resistance of the ether.

Another hypothesis occurred, from which an explanation was attempted of this and of some great inequalities in the motions of Jupiter and Saturn, that seemed not to return periodically, and were therefore nearly in the same circumstances with the moon's acceleration. It was observed, that most of the agents we are acquainted with, take time to pass from one point of space to another ; that the force of gravity may be of this sort, and may not, any more than light, be instantaneously transmitted from the sun to the planets, or from the planets to one another. The effect that would

arise from the time thus taken up by gravity, in
its transmission from one point of space to another,
was therefore investigated by the strictest laws of
geometry ; but it was found, that this hypothesis
did not, any more than the preceding, afford an
explanation of the moon's acceleration.

By this time, also, it was demonstrated, that
there was not, and could not be, in our system,
any inequality whatever produced by the mutual
gravitation of the planets, that was not periodical,
and that did not, after reaching a certain extent,
go on to diminish by the same law that it had increased.

An entire suspense of opinion concerning the
moon's acceleration therefore took place, till Laplace found out a truth that had eluded the search
of every other mathematician.   It was known to
him, both from the investigation of Lagrange, and
from his own, that there are changes in the eccentricities of the planetary orbits, extremely slow,
and of which the full series is not accomplished but
in a very long period.   The eccentricity of the
earth's orbit is subject to this sort of change ; and
as some of the lunar inequalities are known to depend on that eccentricity, they must vary slowly a·
long with it ; and hence an irregularity of a very
long period in the moon's motion.   On examining
further, and the examination was a matter of great
difficulty, Laplace found this inequality to answer

very exactly to what we have called the acceleration of the moon ; for though, in strictness, it is
not uniform, it varies so slowly, that it may be accounted uniform for all the time that astronomical
observation has yet existed.   It is a quantity of
such a kind, and its period of change is so long,
that for an interval of two thousand years, it may
be considered as varying uniformly.   Two thousand years are little more than an infinitesimal in
this reckoning ; and as an astronomer thinks he
commits no error when he considers the rate of the
sun's motion as uniform for twenty-four hours, so
he commits none when he regards the rate of this
equation as continuing the same for twenty centuries.   That man, whose life, nay, the history of
whose species, occupies such a mere point in the
duration of the world, should come to the knowledge of laws that embrace myriads of ages in their
revolution, is perhaps the most astonishing fact
that the history of science exhibits.

Thus Laplace put the last hand to the theory of
the moon, nearly one hundred years after that
theory had been propounded in the first edition of
the Principia.

The branch of the theory of disturbing forces
that relates to the action of the primary planets on
one another, was cultivated during the same period,
with equal diligence, and with equal success.   In
the years 1748 and 1752, the Academy of Scien-

ces proposed for prize questions the inequalities of Jupiter and Saturn ; both the prizes were gained by Euler, whose researches have thrown so much light on all the more difficult questions, both of the pure and the mixed mathematics. There was a particular difficulty that attended this inquiry, and distinguished it greatly from the case of the moon disturbed in its motion by so distant a body as the sun. In the case of Jupiter and Saturn, the disturbing body may be as near to the one disturbed as this last is to the body about which it revolves ; for the distance of Saturn from Jupiter may sometimes be nearly the same with that of Jupiter from the sun. In such cases, the means of obtaining a series expressing the force of the one planet on the other, and converging quickly, was quite different from any thing required in the case of the moon, and was a matter of extreme difficulty. No man was more fit than Euler to contend with such a difficulty ; he accordingly overcame it ; and his mode of doing so has served as the model for all the similar researches that have since been made. It resulted from his investigation, that both the planets were subject to considerable inequalities, depending on the action of one another, but all of them periodical, and returning after certain stated intervals, not exceeding twenty or thirty years, nearly in the same order.

Though this agreed well with astronomical ob-

servations so far as it went, yet it afforded no account of two inequalities of very long periods, or
perhaps of indefinite extent, which, by the comparison of ancient and modern observations, seemed
to affect the motions of these two planets in opposite directions.

This was a subject, therefore, that remained for
farther discussion. In the mean time, it was considered that the other planets must no doubt be affected in the same way ; and both Euler and
Clairaut gave computations of the disturbance
which the earth suffers from Jupiter, Venus, and
the Moon. The same was extended to the other
planets ; and a great additional degree of accuracy
was thus given to all the tables of the planetary
motions.

In the course of these researches, the change in
the obliquity of the ecliptic came first to be perfectly recognized, and ascribed to the action of the
planets above named on the Earth. It was proved
by Euler, that the change in this obliquity is periodical, like all the others we have already seen ;
that it is not a constant diminution, but a small
and slow oscillation on each side of a mean quantity, by which it alternately increases and diminishes
in the course of periods which are not all of the
same length, but by which, in the course of many
ages, a compensation ultimately takes place.

Still, however, the secular inequalities to which

Jupiter and Saturn were subject, and which seemed
to increase continually without any appearance of
returning into themselves, were not accounted for;
so that the problem of their disturbance was either
imperfectly resolved, or they must be supposed to be
subject to the action of a force different from their
mutual attraction. In the course of about twenty
centuries to which astronomical observation had
extended, it appeared that the motion of Jupiter
had been accelerated by 3° 23′, and that of Saturn
retarded by 5° 13′. This had been first remarked
by Dr Halley, and had been confirmed by the calculations of all the astronomers who came after him.

With a view to explain these appearances, Euler,
resuming the subject, found two inequalities of long
periods that belonged to Jupiter and Saturn; but
they were not such as, either in their quantity or in
their relation to one another, agreed with the facts
observed. Lagrange afterwards undertook the same
investigation; but his results were unsatisfactory;
and Laplace himself, on pushing his approximation
farther than either of the other geometers had done,
found that no alteration in the mean motion could
be produced by the mutual action of these two
planets. Physical astronomy was now embarrassed
with a great difficulty, and at the same time was on
the eve of one of the noblest discoveries it has ever
made. The same Lagrange, struck with this circumstance, that the calculus had never yet given

any inequalities but such as were periodical, applied himself to the study of this general question, whether, in our planetary system, such inequalities as continually increase, or continually diminish, and by that means affect the mean motion of the planets, can ever be produced by their mutual gravitation. He found, by a method peculiar to himself, and independent of any approximation, that the inequalities produced by the mutual action of the planets, must, in effect, be all periodical : that amid all the changes which arise from their mutual action, two things remain perpetually the same ; viz. the length of the greater axis of the ellipse which the planet describes, and its periodical time round the sun, or, which is the same thing, the mean distance of each planet from the sun, and its mean motion remain constant. The plane of the orbit varies, the species of the ellipse and its eccentricity change ; but never, by any means whatever, the greater axis of the ellipse, or the time of the entire revolution of the planet. The discovery of this great principle, which we may consider as the bulwark that secures the stability of our system, and excludes all access to confusion and disorder, must render the name of Lagrange for ever memorable in science, and ever revered by those who delight in the contemplation of whatever is excellent and sublime. After Newton's discovery of the elliptic orbits of the planets, Lagrange's discovery of their periodical inequalities is, without doubt, the noblest truth in physical astronomy ; and, in respect of the doctrine of final causes, it may truly be regarded as the greatest of all.

The discovery of this great truth, however, on the present occasion, did but augment the difficulty with respect to those inequalities of Jupiter and Saturn, that seemed so uniform in their rate ; and it became now more than ever probable, that some extraneous cause, different from gravitation, must necessarily be recognized.

It was here that Laplace stepped in again to extricate philosophers from their dilemma.   On subjecting the problem of the disturbances of the two planets above mentioned, to a new examination, he found that some of the terms expressing the inequalities of these planets, which seemed small, as they involved the third power of the eccentricities, had very long periods, depending on five times the mean motion of Saturn *minus* twice the mean motion of Jupiter, which is an extremely small quantity, the mean motion of Jupiter being to the mean motion of Saturn in a ratio not far from that of five to two.   Hence it appeared, that each of these planets was subject to an inequality, having a period of nine hundred and seventeen years, amounting in the case of the former, when a maximum, to 48′ 44″, and in that of the other to 20′ 49″, with opposite signs.

These two results, therefore, are deduced from the theory of gravitation, and, when applied to the comparison of the ancient and modern observations, are found to reconcile them precisely with one ano-- ther. The two equations had reached their maximum in 1560 : from that time, the apparent mean motions of the planets have been approaching to the true, and became equal to them in 1790. Laplace has further observed, that the mean motions which any system of astronomy assigns to Jupiter and Saturn, give us some information concerning the time when that system was formed. Thus, the Hindoos seem to have formed their system when the mean motion of Jupiter was the slowest, and that of Saturn the most rapid ; and the two periods which fulfil these conditions, come very near to the year 3102 before the Christian era, and the year 1491 after it, both of them remarkable epochs in the astronomy of Hindostan.

Thus, a perfect conformity is established between theory and observation, in all that respects the disturbances of the primary planets and of the moon ; there does not remain a single inequality unexplained ; and a knowledge is obtained of several, of which the existence was indicated, though the law could not have been discovered by observation alone.

The discoveries of Laplace had first been communicated in the Memoirs of the Academy of

Sciences; as those of the other mathematicians above mentioned had been, either in these same Memoirs, or in those of Petersburgh and Berlin. An important service is rendered to science, by bringing all these investigations into one view, as is done in the Mécanique Céleste, and deducing them from the same principles in one and the same method. Laplace, though far from the only one who had signalized himself in this great road of discovery, being the person who had put the last hand to every part, and had overcome the difficulties which had resisted the efforts of all the rest, was the man best qualified for this work, and best entitled to the honour that was to result from it. Indeed, of all the great co-operators in this unexampled career of discovery, Lagrange and Laplace himself were the only survivors when this work was published.

We cannot dismiss the general consideration of the problem of the Three Bodies, and of the Second Book of the Mécanique Céleste, without taking notice of another conclusion that relates particularly to the stability of the planetary system. The orbits of the planets are all ellipses, as is well known, having the sun in their common focus; and the distance of the focus from the centre of the ellipsis, is what astronomers call the eccentricity of the orbit. In all the planetary orbits, this eccentricity is small, and the ellipse approaches nearly to a circle.

These eccentricities, however, continually change, though very slowly, in the progress of time, but in such a manner that none of them can ever become very great. They may vanish, or become nothing, when the orbit will be exactly circular ; in which state, however, it will not continue, but change in the course of time, into an ellipsis, of an eccentricity that will vary as before, so as never to exceed a certain limit. What this limit is for each individual planet, would be difficult to determine, the expression of the variable eccentricities being necessarily very complex. But, notwithstanding of this, a general theorem, which shows that none of them can ever become great, is the result of one of Laplace's investigations. It is this—If the mass of each planet be multiplied into the square of the eccentricity of its orbit, and this product into the square root of the axis of the same orbit, the sum of all these quantities, when they are added together, will remain for ever the same. This sum is a constant magnitude, which the mutual action of the planets cannot change, and which nature preserves free from alteration. Hence no one of the eccentricities can ever increase to a great magnitude ; for as the mass of each planet is given, and also its axis, the square of the eccentricity in each, is multiplied into a given coefficient, and the sum of all the products so formed, is incapable of change. Here, therefore, we have again another

general property, by which the stability of our system is maintained ; by which every great alteration
is excluded ; and the whole made to oscillate, as it
were, about a certain mean quantity, from which it
can never greatly depart.

If it be asked, Is this quantity necessarily and
unavoidably permanent in all systems that can be
imagined, or under every possible constitution of
the planetary orbits ? We answer, By no means :
if the planets did not all move one way,—if their
orbits were not all nearly circular, and if their
eccentricities were not small, the permanence of the
preceding quantity would not take place. It is a
permanence, then, which depends on conditions
that are not necessary in themselves ; and therefore
we are authorized to consider such permanence as
an argument of design in the construction of the
universe.

When we thus obtain a limit, beyond which all
the changes that can ever happen in our system
shall never pass, we may be said to penetrate, not
merely into the remotest ages of futurity, but to
look beyond them, and to perceive an object, situated, if we may use the expression, on the other
side of infinite duration.

Though, in the detail into which we have now
entered, we have anticipated many things that may
be thought to belong to another place, we think
that the leading facts are in this way least separat-

ed from one another. Laplace, after treating of the problems of the Three Bodies generally in the Second Book, to which the observations made above chiefly refer, resumes the consideration of the same problem; and the application of it to the tables of the planets in the Sixth and Seventh. These we shall be able to pass over slightly, as much of what might be said concerning them is contained in the preceding remarks. We go on now to the Second Volume, which treats of the Figure of the Planets, and of the Tides.

In the First Book, a foundation was laid for this research by the general theorems that were investigated concerning the equilibrium of fluids and the rotation of bodies. These are applied here ; first, to the figure of the planets in general ; and afterwards particularly to the figure of the earth.

The first inquiry into the physical causes which determine the figure of the earth and of the other planets, was the work of Newton, who showed, that a fluid mass revolving on its axis, and its particles gravitating to one another with forces inversely as the squares of their distances, must assume the figure of an oblate spheroid ; and that, in the case of a homogeneous body, where the centrifugal force bore the same ratio to the force of gravity that obtains at the surface of the earth, the equatorial diameter of the spheroid must be to the polar axis as 231 to 230. The method by which this conclusion was deduced was, however, by no means unexceptionable, as it took for granted, that the spheroid must be elliptical. The defects of the investigation were first supplied by Maclaurin, who treated the subject of the Figure of the Earth in a manner alike estimable for its accuracy and its elegance. His demonstration had the imperfection, at least in a certain degree, of being synthetical ; and this was remedied by Clairaut; who, in a book on the Figure of the Earth, treated the subject still more fully ; simplified the view of the equilibrium that determines the figure ; and showed the true connection between the compression at the poles and the diminution of gravity on going from the poles to the equator, whatever be the internal structure of the spheroid. Several mathematicians considered the same subject afterwards ; and, in particular, Legendre proved, that, for every fluid mass given in magnitude, and revolving on its axis in a given time, there are two elliptic spheroids that answer the conditions of equilibrium : in the instance of the Earth, one of these has its eccentricity in the ratio of 231 to 230 ; the other, in the ratio of 680 to 1.

The results of those investigations, with the addition of several quite new, are brought together in the work before us, and deduced according to the peculiar methods of the author. These theoretical conclusions are next applied to the experiments and

observations that have been actually made, whether
by determining the length of the seconds pendulum in different latitudes, or by the measurement
of degrees. After a very full discussion, and a
comparison of several different arches, on each of
which an error is allowed, and this condition super-
added, that the sum of the positive and negative errors shall be equal, and, at the same time, the sum
of all the errors, supposing them positive, shall be
a minimum, Laplace finds that the result is not re-
concileable with the hypothesis of an elliptic spheroid,
unless a greater error be admitted in some of the
degrees than is consistent with probability. In this
determination, however, the Lapland degree is tak-
en as measured by Maupertuis, and the other academicians who assisted him. The correction by
the Swedish mathematicians was not made when
this part of Laplace's work was published. If that
correction is attended to, the result will come out
more favourable to the elliptic theory than he supposes. There are, however, even after that correction is admitted, considerable deviations from the
elliptic figure, such as the attraction of mountains
is hardly sufficient to explain. The degrees that
have been lately measured in France with so much
exactness, compared with one another, give an ellipticity of about $\frac{1}{150}$, and the same ellipticity corresponds well to the degrees measured in the trigonometrical survey of England, whether of the meridi-

an, or the perpendicular to it. At the same time, the measures in France compared with those in Peru, give $\frac{1}{334}$ for the ellipticity of the meridian, which is less than half the former quantity. The observations of the lengths of the pendulum give the same nearly ; so that this may be taken as the mean result.

The Fourth Book of the Mécanique Céleste, treats of the Tides ;—a subject on which much new light has been thrown by the investigations of Laplace.

The first satisfactory explanation which was given of the flux and reflux of the sea, was that of Newton, founded on the principle of attraction. The force of the moon acting on the terrestrial spheroid, supposing this last to be covered with water, must tend, as Newton demonstrated, to diminish the gravity of the waters toward the earth, both at the point where the moon was vertical, and at the point diametrically opposite ; and this is such a ratio, that the waters would assume the figure of an oblong elliptic spheroid, with its greater axis directed to the moon. The sun must affect the great mass of the waters in a similar manner, and produce an aqueous spheroid, that at the time of new and full moon would coincide with the former, and therefore augment its effect ; while at the quarters it would be at right angles to it, and in part destroy that effect.

The subject, however, was not so fully handled by Newton, but that great room appeared for improvements; and accordingly, the subject of the tides was proposed as the prize-question by the Academy of Sciences in the year 1740. This produced the three excellent dissertations of Daniel Bernoulli, Euler, and Maclaurin, which shared the prize; but shared it, we must confess, with another essay, that of Father Cavalieri, a Jesuit, who endeavoured to explain the tides by the system of vortices. It is the last time that the vortices entered the lists with the theory of gravitation.

Many excellent dissertations on the same subject have appeared since; but they are all defective in this, that they suppose the waters of the ocean in a state of equilibrium, or to be brought, by the action of gravitation, toward the earth, and toward the two other bodies just mentioned, into the figure of an aqueous spheroid, where the particles of the water, by the action of these different forces, were maintained at rest.

This, however, is by means the case: the rotation of the earth does not allow time to this spheroid ever to be accurately formed; and, long before the three attractions are able to produce their full effect, they are changed relatively to one another, and disposed to produce a different effect. Instead, therefore, of the actual formation of an aqueous spheroid, the tendency to it produces a continual oscillation in the waters of the ocean, which are thus preserved in perpetual movement, and never can attain a state of equilibrium and of rest. To determine the nature of these oscillations, however, is a matter of extreme difficulty, and is a problem which neither Newton, nor any of the three geometers who pursued his track, was able, in the state of mechanical and mathematical science which then existed, to resolve. The best thing which they could do, was that which they actually accomplished, by inquiring into the nature of the spheroid, which, though never actually attained, was an ideal mean to which the real state of the waters made a periodical and imperfect approach. Neither the state of mechanical or mathematical science was such as could yet enable any one to determine the motions of a fluid, acted on by the three gravitations above mentioned, and having, besides, a rotatory motion. The nature of fluids was not so well known as to admit of the differential equations containing the conditions of such motions to be exhibited ; and mathematical science was not so improved as to be able to integrate such equations. The first man who felt himself in possession of all the principles required to this arduous investigation, and who was bold enough to undertake a work, which, with all these resources, could not fail to involve much difficulty, was Laplace ; who, in the years 1775, 1779, and 1790,

communicated to the Academy of Sciences a series
of memoirs on this subject, which he has united
and extended in the fourth book of the Mécanique
Celeste.

Considering each particle of water as acted on by
three forces, its gravitation to the earth, to the sun
and to the moon, and also as impelled by the rotation
of the earth, he inquires into the nature of the oscillations that will be excited in the fluid.   He
finds, that the oscillations thus arising may be divided into three classes.   The first do not depend
on the rotation of the earth, but only on the motion of the sun or moon in their respective orbits,
and on the place of the moon's nodes.   These oscillations vary periodically, but slowly; so that
they do not return in the same order, till after a
very long interval of time.   The oscillations of the
second class, depend principally on the rotation of
the earth, and return in the same order, after the
interval of a day nearly.   The oscillations of the
third class, depend on an angle that is double the
angular rotation of the earth; so that they return
after the interval nearly of half a day.   Each of
these classes of oscillations, proceeds just as if the
other two had no existence; a circumstance that
tends very much to simplify the investigation into
their combined effect.

The oscillations of the first kind are proved to
be almost entirely destroyed by the resistance which

any motion of the whole sea must necessarily meet with ; and they amount nearly to the same as if the sea were reduced at every instant to an equilibrium under the attracting body.

The oscillations of the second class involve, in the expression of them, the rotation of the earth ; and they are also affected by the depth of the sea. The difference of the two tides in the same day, depends chiefly on these oscillations ; and it is from thence that Laplace determines the mean depth of the sea to be about four leagues.

The oscillations of the third kind, are calculated in the same manner ; and from the combination of all these circumstances, the height of the tides in different latitudes, in different situations of the sun and moon,—the difference between the consecutive tides,—the difference between the time of high water and the times when the sun and moon come to the meridian,—all these circumstances are better explained in this method than they have ever been by any other theory. Laplace has instituted a very elaborate comparison between his theory and observations on the tides, made during a succession of years at Brest, a situation remarkably favourable for such observations.

1. Between the laws by which the tides diminish from their maximum at the full and change, to their minimum at the first and third quarters, and by which they increase again from the mini-

mum to the maximum, as deduced from the observations at Brest, and as determined by the theory
of gravitation, there is an exact coincidence.

2. According to theory, the height of the tides,
at their maximum, near the equinoxes, is to their
height in similar circumstances at the solstices,
nearly as the square of the radius to the square of
the cosine of the declination of the sun at the sol-
stice ; and this is found to agree nearly with observation.

3. The influence of the moon on the tides increases as the cube of her parallax ; and this agrees
so well with observation, that the law might have
been deduced from observation alone.

4. The retardation of the tides from one day to
another, is but half as great at the syzigies as at the
quadratures. This is the conclusion from theory ;
and it agrees well with observation, which makes
the daily retardation of the tide 27' in the one
case, and 55' in the other.

Many more examples of this agreement are mentioned ; and it is highly satisfactory to find the genuine results of the theory of gravitation, when deduced with an attention to all the circumstances,
and without any hypothetical simplification whatsoever, so fully confirmed in the instance that is
nearest to us, and the most obvious to our senses.

Laplace has treated a subject connected with the
tides, that, so far as we know, has not been touch-

ed on by any author before him. This is the stability of the equilibrium of the sea. A fluid surrounding a solid nucleus, may either be so attracted to that nucleus, that, when any motion is communicated to it, it will oscillate backwards and forwards till its motion is destroyed by the resistance it meets with, when it will again settle into rest; or it may be in such a state, that when any motion is communicated to it, its vibrations may increase, and become of enormous magnitude. Whether the sea may not, by such means, have risen above the tops of the highest mountains, deserves to be considered ; as that hypothesis, were it found to be consistent with the laws of nature, would serve to explain many of the phenomena of natural history. Laplace, with this view, has inquired into the nature of the equilibrium of the sea, or into the possibility of such vast undulations being propagated through it. The result is, that the equilibrium of the sea must be stable, and its oscillations continually tending to diminish, if the density of its waters be less than the mean density of the earth ; and that its equilibrium does not admit of subversion, unless the mean density of the earth was equal to that of water, or less. As we know, from the experiments made on the attraction of mountains, as well as from other facts, that the sea is more than four times less dense than the materials which compose the solid nucleus of the globe are at a medium, the possibility of these great undulations is entirely excluded ; and therefore, says Laplace, if, as cannot well be questioned, the sea has formerly covered continents that are now much elevated above its level, the cause must be sought for elsewhere than in the instability of its equilibrium.

With the questions of the figure of the earth, and of the flux and reflux of the sea, that of the precession of the equinoxes is closely connected ; and Laplace has devoted his fifth book to the consideration of it. This motion, though slow, being always in the same direction, and therefore continually accumulating, had early been remarked, and was the first of the celestial appearances that suggested the idea of an *annus magnus,* one of those great astronomical periods by which so many days and years are circumscribed. As it affects the whole heavens, and as the changes it produces are spread out over the vast extent of 25,000 years, it has proved a valuable guide amid the darkness of antiquity, and has enabled the astronomer to steer his course with tolerable certainty, and here and there to discover a truth in the midst of the traditions and fables of the heroic ages.

Newton was the first who turned his thoughts to the physical cause of this appearance ; and it required all the sagacity and penetration of that great man to discover this cause in the principle of universal gravitation. The effect of the forces of the

sun and moon on that excess of matter which sur-
rounds the earth at the equator, must, as he has
proved, produce a slow angular motion in the plane
of the latter, and in a direction contrary to that of
the earth's rotation.   The accurate analysis of the
complicated effect that was thus produced, was a
work that surpassed the power, either of geometry
or mechanics, at the time when Newton wrote ;
and his investigation, accordingly, was founded on
assumptions that, though not destitute of probability, could not be shown to be perfectly conformable to truth ; and it even involved a mechanical
principle, which was taken up without due consideration.   Nevertheless, the glory of having been
first in the career, is not tarnished by a partial failure, and is a possession which the justice of posterity does not suffer Newton to share even with
those who have since been more successful in their
researches.

    The first of these was D'Alembert.   That ex-
cellent mathematician gave a solution of this problem that has never been surpassed for accuracy
and depth of reasoning, though it may have been,
for simplicity and shortness.   He employed the
principle already ascribed to him of the equilibrium
among the forces destroyed when any change of
motion is produced ; and it was by means of the
equations that this proposition furnished, that he
was enabled to proceed without the introduction of

hypothesis. Solutions of the same problem have since been given by several mathematicians, by Thomas Simpson, Frisi, Walmsley, &c. and many others ; not, however, without some difference (such as the difficulty of the investigation) in the results they have obtained. Laplace has gone over the same ground, more that he might give unity and completeness to his work, than that he could expect to add much to the solution of D'Alembert. As he has proceeded in a more general manner than the latter, he has obtained some conclusions not included in this solution. He has shown, that the phenomena of the precession and nutation must be the same in the actual state of our terraqueous spheroid, as if the whole was a solid mass ; and that this is true, whatever be the irregularity of the depth of the sea. He shows also, that currents in the sea, rivers, trade-winds, even earthquakes, can have no effect in altering the earth's rotation on its axis. The conclusions with regard to the constitution of the earth that are found to agree with the actual quantity of the precession of the equinoxes are, that the density of the earth increases from the circumference toward the centre ; that it has the form of an ellipsoid of revolution, or, as we use to call it, of an elliptic spheroid ; and that the compression of this spheroid at the poles is between the limits of $\frac{1}{304}$ and $\frac{1}{578}$ part of the radius of the equator.

The second part of Laplace's work has for its

object, a fuller developement of the disturbances of
the planets, both primary and secondary, than was
compatible with the limits of the first part. After
the ample detail into which we have entered concerning two of these subjects, the theory of the
moon, and the perturbations of the primary planets,
we need not enlarge on them further, though they
are prosecuted in the second part of this work,
and form the subject of the sixth and seventh
books. In the second book, the inequalities had
been explained, that depend on the simple power
of the eccentricity: here we have those that depend on the second and higher powers of the same
quantity; and such are the secular equations of
Jupiter and Saturn, above mentioned. The nu-
meral computations are then performed, and every
thing prepared for the complete construction of
astronomical tables, as the final result of all these
investigations. The calculations, of course, are of
vast extent and difficulty, and incredibly laborious.
In carrying them on, Laplace had the assistance,
as he informs us, of Delambre, Bouvard, and other
members of the Institute. The labour is, indeed,
quite beyond the power of any individual to execute.

The same may be said of the seventh book,
which is devoted to a similar developement of the
lunar theory. We can enter into no further detail on this subject. One fact we cannot help
mentioning, which is to the credit of two British
astronomers, Messrs Mason and Dixon, who gave

a new edition of Mayer's tables, more diligently
compared with observation, and therefore more accurate, than the original one. Among other improvements, was an *empirical* equation, amounting
to a little more than 20″ when a maximum, which
was not founded on theory, but was employed because it made the tables agree better with observation. As this equation, however, was not derived
from principle (for the two astronomers just named,
though accurate observers and calculators, were
not skilled enough in the mathematics, to attempt
deducing it from principle) it was generally rejected by other astronomers. Laplace, however,
found that it was not to be rejected ; but, in reality, proceeded from the compression of the earth at
the poles, which prevents the gravitation to the
earth from decreasing, precisely as the squares of
the distances increase, and by that means produces
this small irregularity. The quantity of the polar
compression that agrees best with this, and some
other of the lunar irregularities, is nearly that
which was stated above, $\frac{1}{304}$ of the mean radius of
the earth. The ellipticity of the sun does, in like
manner, affect the primary planets ; but, as its influence diminishes fast as the distance increases, it
extends no further (in any sensible degree) than
the orbit of Mercury, where its only effect is to
produce a very small direct movement of the line
of the apsides, and an equal retrograde motion of

the nodes, relatively to the sun's equator. We may judge from this, to what minuteness the researches of this author have extended : and, in general, when accuracy is the object to be obtained, the smaller the quantity to be determined, the more difficult the investigation.

The eighth book has for its object, to calculate the disturbances produced by the action of the secondary planets on one another ; and particularly refers to the satellites of Jupiter, the only system of secondary planets on which accurate observations have been, or, probably, can be made. Though these satellites have been known only since the invention of the telescope, yet the quickness of their revolutions has, in the space of two centuries, exhibited all the changes which time developes so slowly in the system of the primary planets ; so that there are abundant materials for a comparison between fact and theory. The general principles of the theory are the same that were explained in the second book ; but there are some peculiarities, that arise from the constitution of Jupiter's system, that deserve to be considered. We have seen above, what is the effect of commensurability, or an approach to it, in the mean motion of contiguous planets ; and here we have another example of the same. The mean motions of the three first satellites of Jupiter, are nearly as the numbers 4, 2, and 1 ; and hence a periodical system of inequali-

ties, which our astronomer Bradley was sharp-
sighted enough to discover in the observation of
the eclipses of these satellites, and to state as
amounting to 437.6 days. This is now fully explained from the theory of the action of the satellites.

Another singularity in this secondary system, is,
that the mean longitude of the first satellite *minus*
three times that of the second, *plus* twice that of
the third, never differs from two right angles, but
by a quantity almost insensible.

One can hardly suppose that the original motions were so adjusted, as to answer exactly to this
condition ; it is more natural to suppose that they
were only nearly so adjusted, and that the exact
coincidence has been brought about by their mutual action. This conjecture is verified by the
theory, where it is demonstrated that such a change
might have been actually produced in the mean
motion by the mutual action of those planetary
bodies, after which the system would remain stable,
and no further change in those motions would
take place.

Not only are the mutual actions of the satellites
taken into account in the estimate of their irregularities, but the effect of Jupiter's spheroidal figure
is also introduced. Even the masses of the satellites are inferred from their effect in disturbing the
motions of one another.

In the ninth book Laplace treats of Comets, of
the methods of determining their orbits, and of
the disturbances they suffer from the planets.
We cannot follow him in this; and have only to
add, that his profound and elaborate researches are
such as we might expect from the author of the
preceding investigations.

The tenth book is more miscellaneous than any
of the preceding; it treats of different points relative to the system of the world. One of the most
important of these is astronomical refraction. The
rays of light from the celestial bodies, on entering
the earth's atmosphere, meet with strata that are
more dense the nearer they approach to the earth's
surface; they are, therefore, bent continually toward the denser medium, and describe curves that
have their concavity turned toward the earth. The
angle formed by the original direction of the ray,
and its direction at the point where it enters the
eye, is called the astronomical refraction. Laplace
seeks to determine this angle by tracing the path
of the ray through the atmosphere; a research of
no inconsiderable difficulty, and in which the
author has occasion to display his skill both in
mathematical and in inductive investigation. The
method he pursues in the latter, is deserving of attention, as it is particularly well adapted to cases
that occur often in the more intricate kinds of physical discussion.

4

The path of the ray would be determined from the laws of refraction, did we know the law by which the density of the air decreases from the earth upwards. This last, however, is not known, except for a small extent near the surface of the earth, so that we appear here to be left without sufficient data for continuing the investigation. We must, therefore, either abandon the problem altogether, or resolve it hypothetically, that is, by assuming some hypothesis as to the decrease of the density of the atmosphere. Little would be gained by this last, except as an exercise in mathematical investigation, if it were not that the total quantity of the refraction for a given altitude can be accurately determined by observation. Laplace, availing himself of this consideration, begins with making a supposition concerning the law of the density, that is not very remote from the truth, (as we are assured of from the relation between the density of air and the force with which it is compressed ;) and he compares the horizontal refraction, calculated on this assumption, with that which is known to be its true quantity. The first hypothesis which he assumes, is that of the density being the same throughout : this gives the total refraction too small, and falls on that account to be rejected, even if it were liable to no other objection. The second hypothesis supposes a uniform temperature through the whole extent of the atmosphere, or it supposes that the density decreases in geometrical proportion, while the distance from the earth increases in arithmetical. The refraction which results is too great, so that this supposition must also be rejected.

If we now suppose the density of the air to decrease in arithmetical progression, while the height does the same, and integrate the differential equation to the curve described by the ray; on this hypothesis, the horizontal refraction is too small, but nearer the truth than on the first hypothesis. A supposition intermediate between that which gave the refraction too great, and this which gives it too small, is therefore to be assumed as that which approaches the nearest to the truth. It is this way of limiting his conjectures by repeated trials, and of extracting from each, by means of the calculus, all the consequences involved in it, that we would recommend to experimenters, as affording one of the most valuable and legitimate uses of hypothetical reasoning. He then employs an intermediate hypothesis for the diminution of the density of the air ; which it is not easy to express in words ; but from which he obtains a result that agrees with the horizontal refraction, and from which, of course, he proceeds to deduce the refraction for all other altitudes. The table, so constructed, we have no doubt, will be found to contribute materially to the accuracy of astronomical observation.

The researches which immediately follow this, relate to the terrestrial refraction, and the measurement of heights by the barometer. The formula given for the latter, is more complicated than that which is usually employed with us in Britain, where this subject has been studied with great care. In one respect, it is more general than any of our formulas; it contains an allowance for the difference of latitude. We are not sure whether this correction is of much importance, nor have we had leisure to compare the results with those of General Roy and Sir George Shuckborough. We hardly believe, that, in point of accuracy, the two last can easily be exceeded.

The book concludes with a determination of the masses of the planets, more accurate than had been before given; and even of the satellites of Jupiter. " Of all the attempts of the Newtonian philosophy," says the late Adam Smith in his History of Astronomy, " that which would appear to be the most above the reach of human reason and experience, is the attempt to compute the weights and densities of the sun, and of the several planets." What would this philosopher have said, if he had lived to see the same balance in which the vast body of the sun had been weighed, applied to examine such minute atoms as the satellites of Jupiter?

Such is the work of Laplace, affording an example, which is yet solitary in the history of human knowledge, of a theory entirely complete ; one that has not only accounted for all the phenomena that were known, but that has discovered many before unknown, which observation has since recognized. In this theory, not only the elliptic motion of the planets, relatively to the sun, but the irregularities produced by their mutual action, whether of the primary on the primary, of the primary on the secondary, or of the secondary on one another, are all deduced from the principle of gravitation, that mysterious power, which unites the most distant regions of space, and the most remote periods of duration. To this we must add the great truths brought in view and fully demonstrated, by tracing the action of the same power through all its mazes :—That all the inequalities in our system are periodical ; that, by a fixed appointment in nature, they are each destined to revolve in the same order, and between the same limits ; that the mean distances of the planets from the sun, and the time of their revolutions round that body, are susceptible of no change whatsoever ; that our system is thus secured against natural decay,—order and regularity preserved in the midst of so many disturbing causes,—and anarchy and misrule eternally proscribed.

The work where this sublime picture is delineated, does honour, not to the author only, but to the human race ; and marks, undoubtedly, the highest point to which man has yet ascended in the scale of intellectual attainment. The glory, therefore, of having produced this work, belongs, not to the author alone, but must be shared, in various proportions, among the philosophers and mathematicians of all ages. Their efforts, from the age of Euclid and Archimedes, to the time of Newton and Laplace, have all been required to the accomplishment of this great object ; they have been all necessary to form one man for the author, and a few for the readers, of the work before us. Every mathematician who has extended the bounds of his science ; every astronomer who has added to the number of facts, and the accuracy of observation ; every artist who has improved the construction of the instruments of astronomy—all have co-operated in preparing a state of knowledge in which such a book could exist, and in which its merit could be appreciated. They have collected the materials, sharpened the tools, or constructed the engines employed in the great edifice, founded by Newton, and completed by Laplace.

In this estimate we detract nothing from the merit of the author himself; his originality, his invention, and comprehensive views, are above all praise ; nor can any man boast of a higher honour than that the genius of the human race is the only rival of his fame.

This review naturally gives rise to a great varie-

ty of reflections.   We shall state only one or two
of those that most obviously occur.

When we consider the provision made by nature
for the stability and permanence of the planetary
system, a question arises, which was before hinted
at,—whether is this stability necessary or contin-
gent, the effect of an unavoidable or an arbitrary
arrangement ?—If it is the necessary consequence
of conditions which are themselves necessary, we
cannot infer from them the existence of design,
but must content ourselves with admiring them as
simple and beautiful truths, having a necessary and
independent existence.   If, on the other hand, the
conditions from which this stability arises necessarily, are not necessary themselves, but the consequences of an arrangement that might have been
different, we are then entitled to conclude, that it
is the effect of wise design exercised in the construction of the universe.

Now, the investigations of Laplace enable us to
give a very satisfactory reply to these questions ;
viz. that the conditions essential to the stability of
a system of bodies gravitating mutually to one
another, are by no means necessary, insomuch that
systems can easily be supposed in which no such
stability exists.   The conditions essential to it, are
the movement of the bodies all in one direction—
their having orbits of small eccentricity, or not far
different from circles—and having periods of revo-

lution not commensurable with one another. Now, these conditions are not necessary ; they may easily be supposed different ; any of them might be changed, while the others remained the same. The appointment of such conditions therefore as would necessarily give a stable and permanent character to the system, is not the work of necessity ; and no one will be so absurd as to argue, that it is the work of chance : It is therefore the work of design, or of intention, conducted by wisdom and foresight of the most perfect kind. Thus the discoveries of Lagrange and Laplace lead to a very beautiful extension of the doctrine of *final causes*, the more interesting the greater the objects are to which they relate. This is not taken notice of by Laplace; and that it is not, is the only blemish we have to remark in his admirable work. He may have thought that it was going out of his proper province, for a geometer or a mechanician to occupy himself in such speculations. Perhaps, in strictness, it is so ; but the digression is natural : and when, in any system, we find certain conditions established that are not necessary in themselves, we may be indulged so far as to inquire, whether any explanation of them can be given, and whether, if not referable to a mechanical cause, they may not be ascribed to intelligence.

When we mention, that the small eccentricity of the planetary orbits, and the motion of the planets in the same direction, are essential to the stability of the system, it may naturally occur, that the comets, which obey neither of these laws in their motion, may be supposed to affect that stability, and to occasion irregularities which will not compensate one another. This would, no doubt, be the effect of the comets that pass through our system, were they bodies of great mass, or of great quantity of matter. There are many reasons, however, for supposing them to have very little density ; so that their effect in producing any disturbance of the planets is wholly inconsiderable.

An observation somewhat of the same kind is applicable to the planets lately discovered. They are very small ; and therefore the effect they can have in disturbing the motions of the larger planets is so inconsiderable, that, had they been known to Laplace, (Ceres only was known,) they could have given rise to no change in his conclusions. The circumstance of two of these planets having nearly, if not accurately, the same periodic time, and the same mean distance, may give rise to some curious applications of his theorems. Both these planets may be considerably disturbed by Jupiter, and perhaps by Mars.

Another reflection, of a very different kind from the preceding, must present itself, when we consider the historical details concerning the progress of physical astronomy that have occurred in the foregoing pages. In the list of the mathematicians and philosophers, to whom that science, for the last sixty or seventy years, has been indebted for its improvements, hardly a name from Great Britain falls to be mentioned. What is the reason of this? and how comes it, when such objects were in view, and when so much reputation was to be gained, that the country of Bacon and Newton looked silently on, without taking any share in so noble a contest? In the short view given above, we have hardly mentioned any but the five principal performers; but we might have quoted several others, Fontaine, Lambert, Frisi, Condorcet, Bailly, &c. who contributed their share to bring about the conclusion of the piece. In the list, even so extended, there is no British name. It is true, indeed, that before the period to which we now refer, Maclaurin had pointed out an improvement in the method of treating central forces, that has been of great use in all the investigations that have a reference to that subject. This was the resolution of the forces into others parallel to two or to three axes given in position and at right angles to one another. In the controversy that arose about the motion of the apsides in consequence of Clairaut's deducing from theory only half the quantity that observation had established, as already stated, Simpson and Walmsley took a part; and their essays are allowed to have great merit. The late Dr Matthew

Stewart also treated the same subject with singular skill and success, in his Essay on the Sun's Distance. The same excellent geometer, in his Physical Tracts, has laid down several propositions that had for their object the determination of the Moon's irregularities. His demonstrations, however, are all geometrical; and leave us to regret, that a mathematician of so much originality preferred the elegant methods of the ancient geometry to the more powerful analysis of modern algebra. Beside these, we recollect no other names of our countrymen distinguished in the researches of physical astronomy during this period; and of these none made any attempt toward the solution of the great problems that then occupied the philosophers and mathematicians of the Continent. This is the more remarkable, that the interests of navigation were deeply involved in the question of the lunar theory; so that no motive, which a regard to reputation or to interest could create, was wanting to engage the mathematicians of England in the inquiry. Nothing, therefore, certainly prevented them from engaging in it, but consciousness that, in the knowledge of the higher geometry, they were not on a footing with their brethren on the Continent. This is the conclusion which unavoidably forces itself upon us, and which will be but too well confirmed by looking back to the particulars which we stated in the beginning of this review, as either essential or highly conducive to the improvements in physical astronomy.

The calculus of the sines was not known in England till within these few years. Of the method of partial differences, no mention, we believe, is yet to be found in any English author, much less the application of it to any investigation. The general methods of integrating differential or fluxionary equations, the criterion of integrability, the properties of homogeneous equations, &c. were all of them unknown ; and it could hardly be said, that, in the more difficult parts of the doctrine of Fluxions, any improvement had been made beyond those of the inventor. At the moment when we now write, the treatises of Maclaurin and Simpson, are the best which we have on the fluxionary calculus, though such a vast multitude of improvements have been made by the foreign mathematicians, since the time of their first publication. These are facts, which it is impossible to disguise ; and they are of such extent, that a man may be perfectly acquainted with every thing on mathematical learning that has been written in this country, and may yet find himself stopped at the first page of the works of Euler or D'Alembert. He will be stopped, not from the difference of the fluxionary notation, (a difficulty easily overcome,) nor from the obscurity of these authors, who are both very clear writers, especially the first of them, but from want of knowing the principles and the

methods which they take for granted as known to
every mathematical reader.   If we come to works
of still greater difficulty, such as the Mécanique
Céleste, we will venture to say, that the number of
those in this island, who can read that work with
any tolerable facility, is small indeed.   If we reckon
two or three in London, and the military schools in
its vicinity, the same number at each of the two
English Universities, and perhaps four in Scotland,
we shall not hardly exceed a dozen ; and yet we
are fully persuaded that our reckoning is beyond
the truth.

If any further proof of our inattention to the
higher mathematics, and our unconcern about the
discoveries of our neighbours were required, we
would find it in the commentary on the works of
Newton, that so lately appeared.   Though that
commentary was the work of a man of talents, and
one who, in this country, was accounted a geometer, it contains no information about the recent discoveries to which the Newtonian system has given
rise ; not a word of the problem of the Three
Bodies, of the disturbances of the planetary motions, or of the great contrivance by which these
disturbances are rendered periodical, and the regularity of the system preserved.   The same silence
is observed as to all the improvements in the integral calculus, which it was the duty of a commentator on Newton to have traced to their origin, and

to have connected with the discoveries of his master. If Dr Horsley has not done so, it could only be because he was unacquainted with these improvements, and had never studied the methods by which they have been investigated, or the language in which they are explained.

At the same time that we state these facts as incontrovertible proofs of the inferiority of the English mathematicians to those of the Continent, in the higher department; it is but fair to acknowledge, that a certain degree of mathematical science, and indeed no inconsiderable degree, is perhaps more widely diffused in England, than in any other country of the world. The Ladies' Diary, with several other periodical and popular publications of the same kind, are the best proofs of this assertion. In these, many curious problems, not of the highest order indeed, but still having a considerable degree of difficulty, and far beyond the mere elements of science, are often to be met with; and the great number of ingenious men who take a share in proposing and answering these questions, whom one has never heard of any where else, is not a little surprising. Nothing of the same kind, we believe, is to be found in any other country. The Ladies' Diary has now been continued for more than a century; the poetry, enigmas, &c. which it contains, are in the worst taste possible; and the scraps of literature and philosophy are so childish

or so old-fashioned, that one is very much at a loss
to form a notion of the class of readers to whom
they are addressed. The geometrical part, however, has always been conducted in a superior style ;
the problems proposed have tended to awaken curiosity, and the solutions to convey instruction in a
much better manner than is always to be found in
more splendid publications. If there is a decline,
therefore, or a deficiency in mathematical knowledge in this country, it is not to the genius of the
people, but to some other cause, that it must be attributed.

An attachment to the synthetical methods of the
old geometers, in preference to those that are purely analytical, has often been assigned as the cause
of this inferiority of the English mathematicians
since the time of Newton. This cause is hinted at
by several foreign writers, and we must say that we
think it has had no inconsiderable effect. The example of Newton himself may have been hurtful in
this respect. That great man, influenced by the
prejudices of the times, seems to have thought that
algebra and fluxions might be very properly used
in the investigation of truth, but that they were to
be laid aside when truth was to be communicated,
and synthetical demonstrations, if possible, substituted in their room. This was to embarrass scientific method with a clumsy and ponderous apparatus, and to render its progress indirect and slow in

an incalculable degree.   The controversy that took
place, concerning the invention of the fluxionary
and the differential calculus, tended to confirm
those prejudices, and to alienate the minds· of the
British from the foreign mathematicians, and the
analytical methods which they pursued.   That this
reached beyond the minds of ordinary men, is clear
from the way in which Robins censures Euler and
Bernoulli, chiefly for their love of algebra, while he
ought to have seen that in the very works which he
criticises with so much asperity, things are performed which neither he nor any of his countrymen, at
that time, could have ventured to undertake.

We believe, however, that it is chiefly in the
public institutions of England that we are to seek
for. the cause of the deficiency here referred to, and
particularly in the two great centres from which
knowledge is supposed to radiate over all the rest
of the island.   In one of these, where the dictates
of Aristotle are still listened to as infallible decrees,
and where the infancy of science is mistaken for its
maturity, the mathematical sciences have never
flourished ; and the scholar has no means of advancing beyond the mere elements of geometry.
In the other seminary, the dominion of prejudice is
not equally strong ; and the works of Locke and
Newton are the text from which the prelections
are read.   Mathematical learning is there the great
object of study ; but still we must disapprove of

the method in which this object is pursued. A
certain portion of the works of Newton, or of some
other of the writers who treat of pure or mixed
mathematics in the synthetic method, is prescribed
to the pupil, which the candidate for academical
honours must study day and night. He must
study it, not to learn the spirit of geometry, or to
acquire the δυναμις ευρητικη by which the theorems were discovered, but to know them as a child
does his catechism, by heart, so as to answer readily to certain interrogations. In all this, the invention finds no exercise; the student is confined within narrow limits; his curiosity is not roused; the
spirit of discovery is not awakened. Suppose that
a young man, studying mechanics, is compelled to
get by heart the whole of the heavy and verbose
demonstrations contained in Keil's introduction,
(which we believe is an exercise sometimes prescribed,) what is likely to be the consequence? The
exercise afforded to the understanding by those
demonstrations, may no doubt be improving to the
mind; but as soon as they are well understood, the
natural impulse is to go on; to seek for something
higher; or to think of the application of the theorems demonstrated. If this natural expansion of
the mind is restrained, if the student is forced to
fall back, and to go again and again over the same
ground, disgust is likely to ensue; the more likely,
indeed, the more he is fitted for a better employ-

ment of his talents ; and the least evil that can be produced, is the loss of the time, and the extinction of the ardour that might have enabled him to attempt investigation himself, and to acquire 'ioth the power and the taste of discovery. Confinement to a regular routine, and moving round and round in the same circle, must, of all things, be the most pernicious to the inventive faculty. The laws of periodical revolution, and of returning continually in the same track, may, as we have seen, be excellently adapted to a planetary system, but are ill calculated to promote the ends of an academical institution. We would wish to see, then, some of those secular accelerations, by which improvements go on increasing from one age to ano· ther. But this has been rarely the case ; and it is melancholy to reflect, how many of the universities of Europe have been the strong holds where prejudice and error made their last stand—the fastnesses from which they were latest of being dislodged. We do not mean to hint that this is true of the university of which we now speak, where the credit of teaching the doctrines of Locke and Newton is sufficient to cover a multitude of sins. Still, however, we must take the liberty to say, that Newton is taught there in the way least conducive to solid mathematical improvement.

Perhaps, too, we might allege, that another public institution, intended for the advancement of science, the Royal Society, has not held out, in the course of the greater part of the last century, sufficient encouragement for mathematical learning. But this would lead to a long disquisition : And we shall put an end to the present digression, with remarking, that though the mathematicians of England have taken no share in the deeper researches of physical astronomy, the observers of that country have discharged their duty better. The observations of Bradley and Maskelyne have been of the utmost importance in this theory ; their accuracy, their number, and their uninterrupted series, have rendered them a fund of immense astronomical riches.   Taken in conjunction with the observations made at Paris, they have furnished Laplace with the *data* for fixing the numerical values of the constant quantities in his different series ; without which, his investigations could have had no practical application.   We may add, that no man has so materially contributed to render the formulas of the mathematician useful to the art of the navigator, as the present astronomer-royal.   He has been the main instrument of bringing down this philosophy from the heavens to the earth ; of adapting it to the uses of the unlearned; and of making the problem of the Three Bodies the surest guide of the mariner in his journey across the ocean.

FINIS.

REVIEW

OF

LE COMPTE RENDU

PAR L'INSTITUT DE FRANCE.

REVIEW

OF

LE COMPTE RENDU

PAR L'INSTITUT DE FRANCE. *

———

AFTER the intercourse of England with the nations of the Continent has been so long and so unhappily interrupted, it cannot but be acceptable to our readers, to receive, from the most enlightened of those nations, an account of the scientific and literary improvements that have taken place in Europe during the last nineteen years. This account is of high authority, consisting of reports made to the Emperor of the French by Committees of the National Institute, about the beginning of the year 1808. These reports, made by command of the Emperor, are mere abstracts or skeletons of more extensive memoirs, which we may expect hereafter to be published. Even the abstracts, however, are interesting ; not only on account of the information they contain, but as belonging to a ceremonial,

* From the Edinburgh Review, Vol. XV. (1809.)—ED.

which, if not quite singular, is certainly very uncommon in the courts of princes.   They are accompanied with very useful notes by the editor
J. L. Kesteloot, a Dutch physician of the University of Leyden.

We are told, that on the 6th of February, his
Majesty being in his Council, a deputation from
the mathematical and physical classes of the National Institute was introduced by the Minister of
the Interior, and admitted to the bar of the Council.   M. Bougainville, the oldest member, and
therefore the president of the class, then addressed
the Emperor in a short speech ; which we shall
give in his own words.

" SIRE,—Votre Majesté Impériale et Royale a
ordonné que les classes de l'Institut viendraient
dans son conseil lui rendre Compte de l'Etat des
Sciences, des Lettres et des Arts, et de leur progrès depuis 1789.

" La classe des Sciences Physiques et Mathématiques s'acquitte aujourd'hui de ce devoir ; et si je
me présente à la tête des savans qui la composent,
c'est à mon âge que je dois cet honneur.

" Mais, SIRE, telle est la diversité des objets
dont cette classe s'occupe, que même avec la pré-
cision dont un savoir profond et l'esprit d'analyse
donnent la faculté, le rapport qui en contient l'ex-
posé exige une grande etendue.

" Ce n'est donc que de l'esquisse, et pour ainsi
10

dire, de la préface de leur ouvrage, que MM. De-lambre et Cuvier vont faire la lecture.

" Je ne me permets qu'une seule observation ; c'est que l'époque de 1789 à 1808, en même temps qu'elle sera pour les événemens politiques et militaires une des plus mémorables dans les fastes des peuples, sera aussi une des plus brillantes dans les annales du monde savant.

" La part qui est due aux Français pour le per-fectionnement des methodes analytiques qui con-duisent aux grandes découvertes du systême du monde, et pour les découvertes même dans les trois régnes de la nature, prouvera que si l'influ-ence d'un seul homme a fait des héros de tous nos guerriers, nos savans, honorés par la protection de votre Majesté qu'ils ont vue dans leurs rangs, sont en droit d'ajouter des rayons à la gloire nation-ale."

After this address from M. Bougainville, which is certainly commendable for its simplicity, though the compliment in the last paragraph might have been better turned, Delambre, secretary of the class of Mathematics, proceeded to read his Re-port, from which we shall select such passages as appear to us the most interesting.

The Report begins with the elementary branches of the mathematics, and takes notice of two trea-tises which have appeared in that department with-in the limits of the period above mentioned,—those

of Legendre and Lacroix. That of Legendre, it
is said, is destined to recal geometry to its ancient
severity, at the same time that it suggests some new
ideas concerning an analytical mode of treating several of the elementary parts of that science. To
understand these two remarks, it must be observed,
that the French mathematicians, having long since
abandoned Euclid, had departed also, in many
things, from the rigour of strict demonstration; a
practice which, in the Elements, where the foundation of the science is to be laid, was surely much
to be condemned. Bossut's Elements of Geometry, which appeared about the year 1775, is almost
the only one in the French language, except the
two here mentioned, where geometrical accuracy
is aimed at throughout. The work of Legendre,
however, has accomplished its object more completely, we think, than that just mentioned, or, indeed, than any other modern treatise of elementary
geometry. It is very extensive, including the properties of the sphere, together with the cubature
and complanation of the solids bounded by planes,
and also of the sphere, cylinder and cone. At the
same time, the propositions contained in it are
purely elementary, that is, such as, by their simplicity and generality, deserve to be considered as
the fundamental truths of the science of geometry.
Among those analytical methods of demonstration,
to which an allusion is made above, we were long

since particularly struck with one, of which, as it happens, we can convey some idea without the assistance of a diagram.

It is well known to those who have compared different treatises of elementary geometry, that one of the greatest difficulties which they present, is the doctrine of parallel lines. Euclid was not able to extricate himself from this difficulty, otherwise than by the introduction of a proposition as an axiom, which certainly is by no means self-evident. Later writers have uniformly experienced the same difficulty; and some of them, trying to avoid the introduction of a new axiom, have fallen into downright paralogisms. Legendre, in his Elements, has given two demonstrations of the properties of parallel lines, without assuming any new axiom. One of these, which is contained in the text, is prolix and less simple than the nature of the theorem to be proved entitles us to expect. The other demonstration, however, which is in the notes, possesses the most perfect simplicity, at the same time that it is new; proceeding on a principle that has been long recognized, but from which no consequence, till now, has ever been deduced.

If we could demonstrate, independently of all consideration of parallel lines, that the three angles of a triangle are together equal to two right angles, the object in view would be accomplished, and the difficulty, in this part of the Elements, would be en-

tirely overcome. Now, the theorem just mentioned would be easily demonstrated, if we had proved, when two angles of one triangle are equal to
two angles of another, that the third angles are also
equal, whatever may be the inequality of the basis,
or of the triangles themselves. Of this proposition, Legendre gives the following demonstration.
If the third angle of a triangle depend not on the
other two angles alone, but on these angles and
also on the base, then is there some *function* of
these angles, and of the base, to which the third
angle is equal. But, if this is true, an equation
exists between the angles of a triangle and one of
its sides; and, if so, a value of that side may be
found in terms of the three angles; that is, the
side has a given ratio to the angles; which is impossible; for they are quantities of different kinds,
and can have no ratio to one another. Whenever,
therefore, two angles of one triangle are equal to
two of another, their third angles are also equal,
whatever their basis may be. This reasoning appears to us extremely ingenious and satisfactory. It
takes for granted nothing but that an angle and a
line are magnitudes which admit of no comparison;
a proposition, of which no one who understands the
terms can entertain the smallest doubt. The reasoning, however, will not be readily followed by those
who are unacquainted with algebra, or to whom the
nature of *functions* and *equations* is not tolerably fa-

miliar.   It is curious, that a principle which all the
world knew, and which was received into geometry
so long ago as the days of Plato, was never made
subservient to the purposes of reasoning, till in the
instance just mentioned, where it is found actually
to involve in it the solution of a great difficulty.
We must, however, take leave of Legendre's treatise, which we cannot sufficiently recommend. The
Elements of Lacroix are also extremely valuable,
though not marked, so strongly as the preceding,
with the characters of originality and invention.

Delambre goes on to remark, that the fine collection of the Greek mathematicians was completed
in 1791, by the Archimedes of Torelli.   We suppose that he has here in view the splendid edition
of Torelli's Archimedes, printed at Oxford, not
indeed in 1791, but in the year following.   He
makes further mention of a translation of the same
into French by M. Peyrard, with a memoir by Delambre himself on the Arithmetic of the Greeks.
" Before this memoir," he adds, " of which your
Majesty yourself condescended to furnish the subject, it was difficult to conceive how the Greeks,
with a notation so imperfect in comparison of ours,
could possibly execute the arithmetical operations
indicated by Archimedes and Ptolemy." This
translation of Archimedes, so far as we know, has
not yet reached England. The memoir of Delambre
must be peculiarly interesting to mathematicians.

On the subject of the ancient geometers and
their writings, we must be indulged in a few more
remarks. What the collection of the Greek geometers is to which Delambre refers, we do not perfectly understand ; but of one thing we are certain, that that collection can never be considered
as complete, while the Collections of Pappus, one
of the most valuable remains of ancient science, are
known only by a very imperfect translation, and
while the original continues shut up in great libra-
ries with other unpublished manuscripts. The
most perfect MS. of Pappus, we believe is at Oxford, and is particularly described by Dr Horsley,
in his restoration of the Inclinations of Apollonius. The late Professor Simson of Glasgow was
the man of all others who had studied Pappus with
the greatest care, as well as the greatest intelligence ; and all the commentaries on that author
which his papers afforded, were deposited in the
Bodleian Library ; so that the University of Oxford is certainly in possession of the best materials
that the world affords, for a correct edition of this
ancient author. We would willingly look to the
learning of that celebrated university for a publication which will be most thankfully received by the
whole mathematical world. *

---

* Though the MSS. of Pappus, we believe, are but few,
there are some now and then to be met with, which an edi-

Before we take leave of that part of the report which relates to the ancient geometry, we must observe, that the most interesting part of it, the geometrical analysis, has not, in later times, been cultivated in France; and very little, as far as we know, in any part of Europe, except Italy and Great Britain. This is so true, that the article of geometrical analysis is not to be found in that great work, which the French philosophers and mathematicians intended as a complete description of the science of the eighteenth century. The neglect, among these philosophers, of a branch of geometry that deserves so well to be cultivated, and is, in fact, one of the most beautiful and elegant inventions in the whole circle of the sciences, is the more wonderful, that the first of the moderns who understood this subject, and who, though destitute of many of the aids which have since been derived from a more complete knowledge of the ancient remains, became a great master of it, was a French geometer. Fermat flourished about the middle of the seventeenth century; and, in his Opera Varia,

---

tor would no doubt think it his duty to consult. One is now in the possession of the Advocates' Library, which was purchased a few years ago. It is very beautifully written; but is probably of no great antiquity. A circumstance that adds to its value is, that the name of Ortous de Mairan is inscribed on it; so that it probably was once the property of that learned and ingenious academician.

has resorted, or re-invented, some of the finest
works of the ancient analysis, and has approached,
at the same time, very near to several of the greatest discoveries of the modern. In the former, however, his course was not followed by the mathematicians of his own country ; and the man who most
nearly trode in the steps of Fermat, was Dr Ed-
mund Halley, in the end of the same century, who,
possessing great learning as well as genius, applied
the former very successfully to the improvement of
science. He was followed by the late Dr Simson
of Glasgow, and Dr Matthew Stewart of Edinburgh, who cultivated the ancient analysis with
singular assiduity ; the former, restoring several of
the most valuable works of the ancients ; and the
latter, introducing the geometrical analysis into
those branches of physical science, which hitherto
had been treated, either in the algebraical manner,
or by synthetical demonstration. The late Dr
Horsley was a proficient in the ancient analysis ;
and we might add some others of this country, who
have cultivated the same subject with success, and
whose writings fall within the period to which the
report of the Institute is limited. In Italy, the
ancient analysis has found several followers; among
the Memoirs of the Società Italiana, many problems are found resolved by it ; but, on the same
subject, we have met with nothing in the Transactions of the other societies of Europe. There must

be something singular in the causes that have promoted the study of a particular branch of science in distant countries, when no concert or peculiar influence can be supposed to have acted exclusively on them.

Delambre insists at some length on the operations in practical geometry, or what the French call Geodesie, that have been lately carried on for the purpose of ascertaining different points relative to the figure of the earth.   The first of the operations to which he refers, is that which was undertaken both by France and England in 1787, for the purpose of ascertaining the distance between the meridian of Paris and that of Greenwich.   He observes, with respect to these, " that, considering the advanced state of the arts and sciences, it was to be expected that the English would endeavour to surpass every thing of the same kind that had yet been done : they succeeded in doing so : the theodolite of Ramsden, the Indian lights used for signals, the new apparatus employed in the measurement of the basis, produced a degree of accuracy hitherto unknown.   The French, on their side, had only angles to measure ; and the repeating circle which Borda had invented, though not of so *imposing* a form as the theodolite of Ramsden, contained in its construction a principle which assured to it a precision at least equal to that

of the latter instrument, and more independent of
the skill of the artist."

We believe, that this encomium on the repeating
circle of Borda is very fairly due to it. That circle puts it in our power, not merely to take a mean
of a great number of observations, but, as those observations are made without being read off till we
come to the last, the error of reading off is no
greater for all the observations put together, than
it would have been for one observation only ; so
that, when divided into as many parts as there have
been observations made, the result almost vanishes.
The repeating circle, therefore, gives a mean of
the errors of observation, and of the division of the
instrument : and the error of reading off, it goes
near to annihilate entirely. This seems to be the
true light in which these instruments should be
viewed ; and as they are now made by Troughton,
with all the accuracy which that excellent artist
gives to whatever passes through his hands, we
should think it highly expedient that a comparison
was instituted between them and the theodolite of
Ramsden, for which the trigonometrical survey affords so good an opportunity.

The success of the measurement of the distance
between the meridians of Greenwich and Paris, led
to the operation on which the new system of measures was founded. The unit fixed on was a qua-

drant of the meridian ; and, under the impossibility of measuring the whole, the largest arch accesible, that between Dunkirk and Barcelona, was chosen. The operations for this purpose began under the direction of Mechain and Delambre, in 1792, and were not concluded till 1799. Of these, Delambre gave an account, in a work that was mentioned in a former Number of this Review. The coincidence of two different bases of 12,000 metres each, and distant 700,000 from one another, demonstrated the extreme accuracy with which the whole had been conducted. Two degrees have been since added, by the continuation of the same meridian to the Balearic Isles.

The same spirit has spread into other parts of Europe ; and has produced important improvements in the science of geography. The astronomer Swanberg measured over again, in 1802, the degree that had been measured in Lapland by Maupertuis, and a party of the French and Swedish academicians. Their measure made the degree of the meridian which cuts the polar circle, to be 57,405 toises,—considerably greater than it was found possible to reconcile, by any theory, with the magnitude of degrees measured in lower latitudes. Melanderjelm, a Swedish astronomer, known by several valuable works, proposed to repeat the measurement ; and the operation was committed to Swanberg and three others, who, using every precaution, and employing the circle of Borda, found the degree, in the latitude of 66° 20′, to be only 57,209 toises; less by 196 toises than the old measurement; agreeing perfectly with other observations; and giving, for the compression at the poles, about one 330th of the earth's semidiameter.

The measurement of Maupertuis and his colleagues occasioned much confusion and debate for near seventy years; and proves, in a remarkable manner, how much worse an inaccurate experiment may often prove than no experiment at all. The great trigonometrical operations carrying on in England are also mentioned by Delambre; though he seems not perfectly informed of their extent. He mentions some in Germany and Switzerland, with which we are not acquainted.

Among the improvements that respect this state of practical geometry, where its operations, by aiming at great accuracy, connect it with more profound and difficult researches, Delambre, with great propriety, reckons the theorem of Legendre, by which the calculation of spherical triangles is reduced, in all the cases of such measurements as we now refer to, to plane trigonometry. The same excellent geometer has extended his theorem to triangles on the surface of a spheroid. (Vid. Mémoire sur les Transcendantes Elliptiques, 1 vol. 4to.)

The enumeration which Delambre gives of the works containing improvements and discoveries in algebra, is very extensive, and includes several treatises which have not yet found their way into this island.   Of those on which we can add, to the short notice which our author gives, some particulars from our own knowledge, we shall select one or two examples.   Lagrange, having accepted the office of professor in the Polytechnic school, composed, for the instruction of his pupils, the work which he calls Calcul des Fonctions, intended as a commentary and supplement to the Théorie des Fonctions Analytiques, which he had before published.   These works are both of great value, on account of the new and accurate view which they give of the principles of differential calculus, or of what we call the method of fluxions.   For many years, the French mathematicians, and indeed the mathematicians of the Continent in general, gave themselves little trouble about the principles of the new geometry ;   and, though they extended its methods, rules, and applications, much beyond what was done in England, they were not so successful in explaining and demonstrating the fundamental truths of the science, as Newton and his followers. This, we believe, will be generally allowed ;   and, till a very late period, scarcely admits of any exception.   Euler himself, though such a master of the Calculus as to have hardly any equal, yet, in the metaphysics, as we may call it, of that calculus, displays none of his usual talent and accuracy of thought.   He contends, that fluxions or differentials have no magnitude whatever, and are truly and literally equal to nothing ; which is a harsh and inaccurate way of stating what is much better expressed by Newton in his doctrine of prime and ultimate ratios ; or by Maclaurin, where he considers fluxions as the measures of velocity.   There were, however, some exceptions to the generality of the observations which we are now making; and D'Alembert, in particular, though he has not written professedly on the subject of the principles of the Calculus, yet, whenever he has occasion to state any thing relative to it, never fails to do so in the luminous manner that we might expect from a geometer who was a metaphysician and a philosopher. Carnot, whose name is so well known, was one of the first among the French mathematicians who treated professedly of the metaphysics of the differential calculus.   The little tract which he wrote on this subject is full of ingenious and sound views ; but such as, though presented in a new form, and one that appeared quite original both to the author and his countrymen, are in reality very little removed from the method of prime and ultimate ratios.   Carnot, however, had the merit of accommodating that method to the form and language of the calculus, better than we were accustomed to do,

4

by stating that a differential equation is not an exact, but only an approximate equation, which comes continually nearer to the truth the less the fluxions or differentials are that are involved in it. Lagrange, however, has placed the matter on the true foundation ; and has shown that, in delivering the general rules of the differential and integral calculus, there is no need for introducing evanescent quantities, or quantities less than any thing that is assigned. Thus, the differential calculus is reduced to the algebra of variable but finite quantities ; and it is only when the application of the general formulas is made to geometric magnitudes, that the ultimate ratios of evanescent quantities come to be considered ; and they do so in a manner that admits of strict demonstration. This step is undoubtedly one of the greatest that has been made in the new analysis since the period of its invention; and we have often wondered that the works of Lagrange, which contain the developement of this idea, have not produced a greater sensation among the mathematicians of this island, who have always aimed so much at accuracy in their manner of treating this subject. We will not allow ourselves to suppose that this proceeds from any illiberal jealousy, or any unwillingness to acknowledge the superior success of a foreigner in a pursuit in which they themselves had been engaged. We must rather ascribe this apparent indifference to the general agitation of Europe, and the interruption of all intercourse, even that of letters, between France and England. On the Continent, these works seem to meet with the reception they deserve. The Théorie des Fonctions was published in the year 5 of the French Revolution. The first edition of the Calcul des Fonctions was in 1805 ; and the second edition, which is now before us, in 1806.

Another treatise of Lagrange is noticed in this report, Traité de la résolution des équations numériques de tous les dégrés.   This is also a work of great merit, and yet it is but little known in this country, though the memoir which is the foundation of it was published by Lagrange in the Berlin Memoirs so long ago as the year 1767.   It deserves to be particularly studied ; and nothing more useful could be done in an elementary treatise of algebra, than to give to this method of approximating to the roots of equations the simplest form which it admits of.

The last article under this head is the Mécanique Céleste of Laplace, on which, as is well known, too much praise cannot be bestowed.   We have already considered this work with a minuteness that renders any further observations on it unnecessary in this place.

The Report mentions three articles in practical mechanics ; the timekeepers for finding the longitude, constructed by Berthoud, which gained the

prize of the Institute ; the hydraulic ram of Mont-
golfier ; and, lastly, a machine approved by the
Class of the Sciences, the Pyreolophorus of Messrs
Lenieps, a new invention, in which, if we understand the very short notice concerning it which the
editor has given in a note, the force of air suddenly expanded by heat, is made to raise a weight, or
overcome a resistance.   In an experiment made
with this machine, it is said, that a boat, loaded
with five quintals, and presenting to the water a
prow of the area of six square feet, was carried up
the Soane with a velocity double that of the stream.
In another experiment, the pressure exerted on a
piston of three square inches was in equilibrio with
21 ounces, and the fuel consumed weighed only
six grains.   We want here a necessary element,
the time in which these six grains were consumed.
This omission may perhaps be supplied from another part of the account, where it appears that each
stroke of the piston takes up five seconds.   The
six grains were the fuel consumed in five seconds.

Much more information, however, than we have
at present, is necessary, in order to form any estimate of the merit of this machine, and to judge
whether it has any chance of becoming a rival to
the steam engine.

The next general head of the Report is Astronomy ; and here the new astronomical tables form
the first, and indeed the most important article.

This subject we have also anticipated in the review of Vince's Astronomy, or, as the title ought to have been, of Vince's edition of the Tables of Burg and Delambre.

A curious article is given with respect to the comet of 1770, which has long occupied the attention of astronomers, from the singular circumstance that the only ellipse that could be made to agree with its motions during the time it was visible, is one in which it must revolve in five years and a half nearly : yet this comet has never been observed since 1770, and never was seen before. The singular problem to which this paradoxical phenomenon gives rise, was proposed as the subject of a prize by the National Institute, and the prize was gained by M. Burckhardt, a most skilful and laborious astronomer. From immense calculations he has made it appear that the attraction of Jupiter had rendered that comet visible, from having been invisible before because of its great distance, and has also rendered it invisible again, by undoing its former effect, and reducing the comet to move in an orbit that does not admit of its coming near enough to the sun to be visible from the earth.

It is not one of the least remarkable circumstances in the history of a period big with novelty, that since the beginning of the present century four new planets have been discovered. These are all of them so small, that it is no wonder they escaped

observation, and were even, by astronomers, con-
founded with the millions of stars of the same apparent magnitude that occupy almost every point
of the heavens. From their smallness it follows,
that they have no sensible effect in disturbing the
motions of the planets already known. Their orbits are considerably eccentric; and the plane of
one of them has an inclination to the ecliptic greater than the inclinations of all the other planetary
orbits put together. This great inclination and
eccentricity will render the calculation of the disturbances produced in the motion of these bodies
by the larger planets, (particularly of Jupiter and
Mars, between which they are situated,) a matter
of considerable difficulty, and may be the occasion,
as Delambre remarks, of extending the science of
analysis beyond its present bounds.

The first of these planets was discovered by
Piazzi at Palermo, the third by M. Harding, the
two others by M. Olbers of Bremen. The astronomer last named is of opinion, that these small
bodies are the fragments of one large planet which
an explosion, from some unknown cause, has burst
in pieces; and hence he concludes, that all their
orbits ought to cut one another in two opposite
points of the heavens, which he found, by calculation, to be, one near the constellation Virgo,
and the other near the Whale; and that, of course,
they must pass through these points twice in every

revolution ; so that, in order to discover all the fragments, astronomers ought to examine these two places of the heavens very frequently.   In effect, all the four have been found near these points; and the two last, after Olbers had suggested the idea just mentioned.

Since the year 1789, seventeen comets have been discovered ; and, along with the names of Messrs Mechain, Olbers, &c. by whom they were observed, we are glad to see the name of Miss Herschel. The orbits of all these comets have been calculated.   The comet that appeared so remarkable in the autumn of 1807, is thought by Olbers to revolve in a very eccentric ellipse, and to have a periodic time of no less than 1900 years.

Delambre concludes this article with Dr Herschel's description of the heavens, the double, triple, quadruple, and nebulous stars, together with those that have disks like planets, in some cases round, in others irregular.   The discovery of the revolution of Saturn's ring by the same excellent astronomer, is also mentioned, as also the coincidence of the time of that revolution with the theory of gravity, and the prediction of Laplace.   The observations of Dr Herschel on the figure of Saturn himself are not mentioned.

A rumour prevailed for some time, that Piazzi had discovered the parallax of the fixed stars ; but as M. Delambre makes no mention of a discovery, which, if real, would be no doubt one of the greatest in astronomy, we must suppose that M. Piazzi's observations have not yet led to any satisfactory result. The notes mention a work, founded mostly on Dr Herschel's observations, by Schrœter of Amsterdam, on the dimensions of the universe : it was crowned by the Royal Society of Haerlem in 1802 ; it cannot fail to be highly interesting; and we very much regret that it has not yet reached this country.

The next general head is that of Physique Mathématique, or what we would call experimental philosophy. Delambre begins with remarking, " That the revolution recently brought about in chemistry, could not happen without turning many experimentalists a little out of their ordinary course, when they saw in a neighbouring science, a road opened that promised more numerous discoveries. We shall nevertheless find, in the mathematical branch of physics, some curious researches and interesting inventions."

Among these, one of the most remarkable is the Balance of Torsion ; which, by the twisting and untwisting of a thread or wire, affords a measure for forces that are too small to be appreciated by any other means. It was with this that Coulomb was so successful in determining the law of electric attraction and repulsion, and afterwards in showing that the phenomena of magnetism follow a law altogether similar, namely, the inverse of the square of the distance. By help of the same instrument, he was able to measure the smallest effects of magnetism ; to find the temperature (considerably elevated) at which these effects entirely disappear ; and to show that magnetism is not, as has been generally supposed, a property peculiar to certain bodies, but one that exists in all, even in those that appear most devoid of it. The same balance enabled him to measure the resistance which fluids oppose to motion, the law of which resistance is expressed by two terms, of which Newton found out only the first, the second not being sensible except in motions that are extremely slow.

The instrument by which Mr Cavendish determined the gravitation of balls of lead toward one another, is, as Delambre observes, no other than Coulomb's Balance, executed on a large scale. Mr Cavendish found from his experiments, that the mean density of the earth is five times and a half as great as that of water.

Here, however, we must be permitted to observe, that though Mr Cavendish's Balance does not differ in principle from that of the excellent experimenter quoted by Delambre, it was not copied from it. The experiments of Mr Cavendish were indeed made about the year 1798; and the first experiments of Coulomb with his balance are published as early as the year 1785, if we mistake not. The instrument that Mr Cavendish employed had, however, been invented before that period by the Rev. Mr Mitchell, F. R. S., and was purchased at his sale by Mr Cavendish. We are to consider the instrument, therefore, as originally an English invention, and re-invented in France by Coulomb, without any knowledge whatever of what was done in England.

We cannot help remarking too, when we reflect on the results obtained from this instrument in the hands of the English and of the French philosopher, that the gravitation measured by the former, may have been affected by the magnetism which the latter supposed he had discovered in all bodies. The effects of the one force may have been mistaken for those of the other, and a degree of uncertainty is thrown on the determinations of both. This observation, however, we only throw out loosely: perhaps an accurate comparison of the experiments might determine how much is to be ascribed to the one cause, and how much to the other: it is right, however, that this source of inaccuracy should be considered.

The application of the barometer to the measurement of heights, or, more properly the formula for determining heights by help of the barometer, deduced by Laplace, is mentioned among the discoveries of the latter. Laplace used in his formula the specific gravity of mercury, as it is commonly stated. The coefficient or multiplier of the logarithmic difference which he thus obtained, was found, on comparing his barometric measures with certain heights, trigonometrically determined by M. Ramond in the Pyrenees, to require a small correction. The coefficient, thus adjusted, was found by Biot to agree perfectly with the experiments on the specific gravity of mercury when accurately repeated ; and his experiments also gave the same refraction which Delambre had deduced from astronomical observations.

In the prosecution of these experiments, M. Biot found that the refracting power of different gases affords means more accurate than the ordinary processes of chemistry for inquiring into the composition of certain substances, such as the diamond, which he concluded to be partly composed of oxygen. The idea of inferring the chemical composition of bodies from their refracting power, as is well known, was first conceived by Newton : it seems to have been much extended and improved on by the philosopher just named.

It is not taken notice of in the Report, but we think it right to remark it, that the rule for barometric measurements had been investigated on strict mathematical and mechanical principles long before it was done by Laplace, and formulas brought out, which do not materially differ in their results, though they do considerably in their forms, from that of the French geometer. After Deluc made his improvements, and discovered by trial the very simple rule which he employs, leaving it however quite empirical, and not deduced from principles, a geometric demonstration of that rule was given by Dr Horsley in the Philosophical Transactions. An investigation of the same, purely analytical, was published by Professor Damen of Leyden; and a third, which considers the problem with great generality, and takes into view several circumstances which had not hitherto been attended to, is given by Professor Playfair in the first volume of the Edinburgh Transactions. The investigation of Laplace, therefore, was not entirely new as to its object or its principles, though we believe his method to be original, and in all respects worthy of its author. His rule, even when corrected as above mentioned, does not perfectly agree with that which we employ in this country, of which the form is agreeable to the investigations just mentioned, and the coefficients determined from the excellent experiments of General Roy and Sir G. Shuckborough. It is also less commodious in practice than either our formula or that of M. Trembley of Geneva. We are not, however, perfectly prepared to state in what the difference consists, or to what extent it goes. As the question now stands, we think a comparison of the different barometric formulas is an excellent subject for a mathematical memoir.

Under the article of Magnetism, the Report mentions the series of observations published by

M. Gilpin in the Philosophical Transactions for 1806, from which some curious results may be deduced concerning the secular variations of the magnetical meridian. Another article relates to Dr Wollaston's apparatus for measuring, in a manner extremely simple and accurate, the refraction of transparent bodies, (Phil. Trans. 1802.) It is said, that a very valuable addition to this apparatus has been made in France, by M. Malus; and that an analytical consideration of the subject had enabled him to correct an error which had escaped Dr Wollaston. We do not know if any more particular account of M. Malus's improvement has yet reached England.

The next object of Delambre's Report, is Geography and Travels. On this he is very short, and only runs over some of the principal occurrences. " The taste," he says, " to which the successful and brilliant voyages of Bougainville and Cook had given rise, was not weakened by the disastrous, though not useless, expeditions of Peyrouse and Entrecasteaux. Deputies from the African Society in England, penetrated into countries entirely unknown. Horneman met with the most distinguished reception from the conqueror of Egypt; Mungo Park braved the greatest dangers; and Flinders exposed himself to the most dreadful risks, in order to explore Van Dieman's Land, and the coast of New Holland. The ambassadors of the

English penetrated into Thibet, into the kingdom of Ava, and into China. Vancouver described the coast he was appointed to survey, with a care and exactness proper to serve as a model for all those who have similar duties to discharge."

We cannot help remarking, on reading the name of Flinders, that the fate of that skilful and intrepid navigator, at this moment, we believe, languishing in confinement in the Isle of Bourbon, does great discredit to the government of France. Accident put him in the power of France. A voyager, engaged in the cause of science, had a right every where to look for friends. Flinders was treated as an enemy. His release, however, was at length agreed to; and orders to that effect sent out to the governor of the Isle of Bourbon: but hitherto, if we are rightly informed, these orders have not been complied with.

The Report goes on to mention what the French did in Egypt; the voyages of Marchand, Baudin, &c.

" Lastly," (says Delambre,) " to terminate this sketch with an expedition which contains in it every kind of merit, Humboldt has executed, at his own expence, an enterprise that would have done honour to a nation. Accompanied only by his friend Bonpland, he has plunged into the American wilderness; he has brought back with him 6000 plants, with their descriptions; has determined the position of 200 points, by astronomical observation;

has ascended to the top, and has measured the height of Chimboraço.   He has created the geography of plants, assigned the limits of vegetation, and of eternal snow ; observed the phenomena of the magnetic needle and of electric fish, and has presented the lovers of antiquity with much valuable information concerning the Mexicans, their language, their history, and monuments."

A sketch of these curious travels is given in one of the notes, at the end of the Report, but would lead us into too long a digression.

Delambre then concludes his Report with a new address to the Emperor.   The Institute had it in command, it seems, not only to report on the actual state of the sciences, but to suggest the measures that would promote their further advancement. In this part of his task, Delambre has acquitted himself well, and with considerable address.

" Votre Majesté daigne interroger l'Institut sur les moyens d'assurer les progrès ultérieurs ; les progrès des mathématiques ne sont nullement douteux, l'instruction première trouve des sources abondantes dans tous les lycées ; l'ecole polytechnique est une pépinière de sujets distingués pour tous les genres de service public.—La loi bienfaisante qui a régeneré l'instruction, promettait une école spéciale aux mathématiques ; cette école existait.   La Géométrie et l'Algébre, l'Astronomie et la Physique sont professées au Collége Impérial de France.

Un cours d'analyse transcendante y complettcrait l'enseignement des sciences exactes.   Une operation importante avait été commencée pour donner à la France une perpendiculaire digne de sa méridienne.——Mais nous ne formons point de vœu; nous attendons avec une confiance respectueuse, ce qu'il plaira à votre Majesté d'ordonner en faveur d'une science dont elle eut elle-même reculé les limites, si des plus hautes destinées ne l'eussent appelée à les protéger toutes également, pour les faire concourir à la splendeur et aux merveilles de son régne."

A school for instruction in the higher mathematics, and a perpendicular to the meridian of Paris, to be extended across the kingdom with the same accuracy as the meridian itself has been, are the very moderate and disinterested requests of the secretary of the Institute, and the things which he conceives to be most essential for promoting the interest of mathematical science.   The respectful manner in which this suggestion is made, and the compliment with which it is accompanied, to some will perhaps appear to savour more of the courtier than the man of science.   We are not, however, of this opinion.   Respect is what talent and power of a certain eminence must always command ; and that a man of the ability of Napoleon, who had early shown a fondness for science, might have enlarged the bounds of it by discoveries of his own, if his situation had permitted, is a natural and fair conclusion.

The report that follows next, is that of Cuvier, on the subject of what we call general physics— *Les Sciences Physiques.* It begins with the theory of crystallisation as given by Haüy, which has originated and been brought to its conclusion, as Cuvier remarks, during the period to which these reports are confined. It is indeed true, that the theory, in the view Haüy takes of it, is completed ; but that the real theory of crystals is understood, till we know the law of the force, whether polarity, or simple attraction, by which the regular structure of these bodies is brought about, we can by no means admit. The cause that arranges the *molecules,*—that determines the rate of the decrease of the different plates of which the crystal is composed,—this is still confessedly unknown ; it is perhaps without our reach ; and if so, we must consider this branch of knowledge as destined to remain for ever imperfect. At the same time, we believe it true, that the principle of Haüy does not admit of being carried much beyond what it has been. We admit, too, that in the state to which it is now brought, it furnishes an excellent principle for the arrangement of mineral substances.

" Within the period we now treat of, the theory of chemical affinities has undergone an entire revolution in the hands of Berthollet, who denies the existence of elective attractions and absolute decomposition. He has undertaken to prove, that in all the compositions and decompositions made by what is called elective attraction, there takes place a division of the substance combined between two other substances that act upon it with opposite forces; and that the proportion in which this division is made, is determined, not only by the absolute energy with which these substances act, but also by their quantity."

It cannot be denied, that, in this way, chemical forces are represented as being more of the nature of those mechanical forces with which we are best acquainted, than in any other. Their nature, therefore, becomes less paradoxical. At the same time, chemists do not seem perfectly agreed as to the solidity of this new theory, and its conformity to the phenomena of their own science. We certainly do not consider ourselves as qualified to decide this question.

In treating of the recent discoveries concerning heat, Cuvier begins with remarking, that they constitute a body of science, of which the philosophers and chemists of the first half of the eighteenth century had not the most distant conception.

" The discoveries of Black in Scotland, and Wilke in Sweden, led the way to this revolution, by showing, not only that a body absorbs a great

quantity of heat when it becomes fluid, and also
when it passes into vapour, which it restores when
it returns to its primitive condition, but also that
unequal quantities of heat are required to increase
the temperature of different bodies by the same
number of degrees. These truths have led to a
great number of others, the influence of which on
the sciences, on the arts, and even the affairs of
common life, is wholly incalculable."

To these discoveries, if we add those of another
kind, in which the same chemists had their share,
the production of fixed air in the burning of charcoal by Mr Cavendish, and of water in the burning of inflammable air by the same philosopher, and
by Monge, and also the augmentation of the
weight of metals by their calcination, and the ab-
sorption of air at the same time, (which last had
been observed by Mr Boyle,) we have the constituent parts of the new chemistry.

" The happiness of uniting all these scattered
rays of knowledge into one pencil, is what consti-
tutes the glory of Lavoisier. Till he appeared,
the particular phenomena of chemistry might be
compared to a kind of labyrinth, of which the deep
and winding paths had been trode by several laborious travellers : but their points of union, their
relation to one another and to the system, could
not be perceived but by the genius which was able
to rise above the edifice, and, with an eagle's eye,

to catch the plan of the whole." Perhaps some will allege, that there is more splendour than solidity in the opinion, which reserves it for the discovery of facts, and withholds all praise from that of their relations. Yet we believe, that, on the whole, this is a fair statement of the merit of Lavoisier. As to what relates to Dr Black, we hope that we are not influenced by national partiality, when we say that Cuvier, not intentionally, (for we think both his report and Delambre's remarkable for their fairness,) has mentioned too slightly the discoveries of our illustrious countryman. His experiments on magnesia were the first that proved the existence of an aeriform fluid becoming fixed in a solid body, and forming an integrant part of it, so considerable as two-fifths of the whole. This was the first step to the creation of pneumatic chemistry.

The new nomenclature of chemistry, and the introduction of a perfectly regular and philosophic language, are next mentioned, as having materially contributed to the advancement of this science. " From all these causes proceed the great things it has accomplished; almost all the substances in nature have been examined; all the imaginable combinations of them exhausted; the number of the metals carried to 28, and of the earths to 9. New acids have been discovered, or have been formed, &c. The names of Berthollet, Fourcroy,

Vauquelin, Chaptal, Guyton, Deyeux, Thenard, among the French ; of Klaproth, Kirwan, Davy, Tennant, and Wollaston, among foreigners, have been rendered immortal."

Speaking of the Galvanic electricity, he observes, that it opens a view into new regions, of which no one can venture to calculate the extent. " The most powerful, perhaps, of all the agents which nature employs in her operations on the surface of the earth, has remained unknown till the present time. We have but just become acquainted with it. The simple juxta-position, not only of two metals, but of two different bodies, whatever they be, alters the equilibrium of the electric virtue ; and this alteration can produce the most violent motions in the animal economy. It separates the substances that are the most closely united. At this moment it seems about to reveal to us the composition of those alkalis, which the most profound chemistry had hitherto regarded as simple bodies. The names of Galvani and Volta, who discovered this mysterious power ; of Ritter, Nicholson, and above all of Davy, who have recognised and found out its chemical action, will occupy a large space in the report we are to make of this new and interesting portion of physical science."

Such is the rapidity, we must observe, with which this part of science is advancing, that Mr Davy has actually accomplished the decomposition of soda and potass, since the time when this report was drawn up ; and has found those alkalis to be no other than oxyds of metallic bodies, hitherto unknown. He has, indeed, found electricity to exercise an absolute command over the most powerful, and, as we supposed, the most simple and independent of chemical agents. These discoveries have procured him the prize offered by the National Institute.

Among the chemical discoveries of the present time, we have been somewhat surprised to find no mention made of that of Sir James Hall, concerning the power of compression to modify the effects of heat. By subjecting limestone to great compression, while heated, the carbonic acid was prevented from escaping ; quicklime was not formed, and the mass was reduced into fusion. This is to be considered as a valuable discovery in chemistry, independently of all the applications of it that may be made to another science. The imperfect communication that takes place between the scientific world of France and England, is probably the cause of this omission.

" Mineralogy now approaches in rigour to the exact sciences; thanks to the crystallographic researches of M. Haüy, to the chemical analyses of Klaproth and Vauquelin, and to the description of the external characters and the position of minerals by Werner and his school.—The description of the relative position of minerals, has now become the object of a real science, and replaces those illusory conjectures which have been called by the name of Geology."

We must observe with respect to this passage, that we entirely agree with what is said on the obligation mineralogy has to Haüy and Werner, and the two chemists mentioned above ; to which chemists several others from this country might easily be added. The Crystallography of Haüy furnishes us with a principle of arrangement that is perfect so far as it extends, and one that defines accurately those *species* into which minerals are divided. This cannot be said of any other system of classification ; not even of Werner's.

As to what concerns Geology, if that science is supposed to treat of the origin of things, or to go back to a period when the composition of the bodies which we call minerals was different from what it is at present, we perfectly agree in thinking that the whole is a most unphilosophical illusion, because maxims founded on our experience of the present order of things cannot apply to what took place before that order was established. But, if by geology is meant an attempt to trace the laws of those changes to which minerals are subject, the changes which they have undergone and are still about to undergo, we see no necessity for its conclusions being illusory and chimerical. Though we

observe minerals only for an instant, or a portion
of time that is quite evanescent, compared with the
great cycles in which the series of their changes
must revolve, yet we may discover such characters
as ascertain important facts in the history of those
changes.   The preliminary investigation is no
doubt that which Cuvier points out,—the relative
situation of the different kinds of minerals, and in
general the accurate description of their present
condition.   But this philosopher does not seem to
be aware, that there is in the very research concerning the present state of minerals as much danger from theory, and from hasty generalisation, as
in the conclusions that geologists have drawn concerning the past or future fortunes of the world.
The language in which Werner and his school
describe the facts concerning the mineral kingdom
is full of theory, and of theory as unsupported and
as remote from experience, as any thing to be met
with in the Cartesian vortices.   The knowledge of
the great facts therefore concerning the relative
position of mineral bodies, though it has made considerable progress, yet, in our opinion, as far as the
observations of the Wernerian school are concerned, is not in that high road to perfection which
this learned and eloquent reporter appears to imagine.   The force that is every day applied to make
the new observations agree with the old, and to as-
similate the structure of the whole world to that of

Saxony and Bohemia, is much more likely to produce retrogradation than advancement.

Our author then passes rapidly over the improvements in physiology, comparative anatomy,
and natural history, and touches on the practical
sciences of medicine and agriculture ; under the
former of which, he particularly mentions vaccina-
tion, and the destruction of contagion by fumiga-
tion.   He goes on to the improvements in the mechanical arts, particularly that of the stereotype
printing, valuable from the cheapness with which
it may be executed ; and thereby promising to
carry the works of genius into the cottage of the
peasant.   We shall only take notice of the conclusion of his report.

" Your Imperial Majesty has commanded this
class to propose the means that seem to it best calculated for maintaining among those who cultivate
the sciences, that emulation by which they are at
present animated ; for directing their efforts to the
most important objects, and for securing to them
successors of equal zeal and ability.

" Without wishing to anticipate the measures
which the wisdom of your Majesty is preparing for
the public education, we will take the liberty, in our
extended report, of submitting some ideas on the
regulation of the first or popular instruction in the
physical sciences, and for spreading more effectually
among the people, the knowledge connected with

husbandry and the useful arts. We have also proposed that your Majesty should ordain the drawing up
of a *new system of physical existences*. Science demands this work ; our country is that in which it
can be most easily executed ; and it would be desirable to see the name of Napoléon, which is already placed at the head of so many great monuments, so many wise laws, and so many useful institutions, decorating the title-page of a fundamental work in science. Of all the establishments
formed, and of all the labours undertaken by the
command of Alexander, Aristotle's History of
Animals is the only one which now remains, an
everlasting testimony of the love of that great
prince for natural knowledge. A word from your
Majesty can create a work which shall as much surpass that of Aristotle in extent, as your actions surpass in splendour those of the Macedonian conqueror."

The answer of the Emperor is very short.

" MM. the Presidents, Secretaries, and De-
    puties of the First Class of the Institute—

" I was desirous to hear you on the progress of
the human mind in these later times, in order that
what you should have to say to me might be heard
by all nations, and might shut the mouths of those
detractors from the present age, who represent
knowledge as retrograde, only because they wish for
its extinction.

" I was also willing to be informed of what remained for me to do to encourage your labours, that
I might console myself for not being able otherwise
to contribute to their success. The welfare of my
people, and the glory of my throne, are equally interested in the prosperity of the sciences.

" My minister of the interior will make a report on your demands. You may constantly rely
on the effects of my protection."

Though we admit that Delambre and Cuvier
have done well; the first, in recommending a
school for instruction in the higher parts of the
mathematics, and an extension of those geodetical
operations from which so much benefit has already
resulted; and the second, in recommending some
further care of the popular instruction in agriculture and the arts, as well as a new and fundamental work on natural history, in its most extensive
sense :—though we are not disposed to quarrel
with the high compliment contained in the prediction, that this work would not farther surpass the
natural history of Aristotle, than the achievements
of Napoleon have exceeded those of Alexander;
yet we are well aware that there are other improvements still more important, and more imperiously
called for, which the spirit of philosophy would demand, if her real and unbiassed sentiments could
be conveyed to the ears of Napoleon. " Cease,"
she would say, " from the fatal and endless pursuit

of military aggrandisement. Give peace to Europe, for your victories enable you to do so ; and let the moderation and liberality of the terms ensure its continuance.   Restore that intercourse and mutual confidence among the nations which are essential to their happiness, no less than to the advancement of knowledge ; and let their prosperity be considered as one of the means of promoting the welfare of your own people. The sciences will then flourish spontaneously, and will require no protection but that which secures their tranquillity and independence ; and you yourself will have the felicity, more singular than all that you have yet experienced, of adding to the titles of Hero and Conqueror, that of the Father of your People."

The National Institute of France is divided into four classes. The first, is that of the mathematical and physical sciences ; the second, that of the French language and literature ; the third, has for its object history and ancient literature ; the fourth, the fine arts. The two reports that we have considered, and which make the principal part of the book before us, are from the first class. The three others are of inferior interest ; and, besides, the length to which our review has already extended precludes our entering on them particularly. In the report from the third class, on the subject of history and ancient literature, speculative philosophy seems, in a certain degree, to be included ;

and we find, accordingly, some notice taken of the revolutions which that philosophy has undergone in Germany and elsewhere. The Ecole d'Ecosse, as the author of the report (M. Lévesque) is pleased to call it, is also made honourable mention of. As no sect of philosophers is known in Scotland by a name which we owe to the politeness of our neighbours, we should have been at some loss to distinguish what system was understood by this phrase, if we had not before met with it in the Histoire Comparée des Systemes de Philosophie, by M. Degerando, where we find this title applied to a succession of philosophers which begins with Dr Hutcheson of Glasgow; comprehends in it Reid, Ferguson, &c. and at present terminates in Professor Dugald Stewart, to whose writings, as Degerando remarks, Reid's philosophy owes its fullest developement, and the greatest share of its celebrity with foreign nations. Sometimes, when the same author speaks more loosely, he appears to include, in the Scottish school, almost all the philosophers that have flourished in that country since the time of Hutcheson, whether they have supported or combated the philosophy of Locke. In this way he includes Lord Kames, David Hume, Adam Smith, &c.; forming a succession of eminent men, of which, in so short a period, and in so narrow a country, there are but few examples in the history of letters.

On the whole, throughout these reports we find great liberality with regard to foreign nations; and if more room is occupied by French improvements and discoveries than by any other, this may be in reality a just allotment; or it may in part be an effect of that perspective which, in intellectual as in visible objects, represents the nearest as the largest, so as sometimes to deceive the justest eye, and the most impartial judgment.

In one instance we think that this fairness is a little departed from, when it is said that no nation has cultivated historical composition so much as the French, or produced so great a number of historians that deserve to be quoted. "It was to a Frenchman," the report adds, " that Italy owed the first history of Rome, written by a modern; and it was a Frenchman who first made the English acquainted with the history of their own country."

Those, however, who have studied history in the best school, will not be very apt to admit, that the dull and unphilosophical narrative of Rapin could bring an Englishman acquainted, as he ought to be, with the history of his country. Whatever the French themselves suppose, it is not the opinion of strangers that they excel in historical composition. For our part, we hope that we are not altogether deceived by national partiality when we say, that we do not know three modern historians,

of any country, that can be compared with three
of which this island boasts, Hume, Robertson, and
Gibbon.   They are historians whom no age but
such a one as the present could produce : it is in
vain to look for any thing ancient to be opposed to
them.   It is only among future generations that
rivals to them can be found.

One remark has struck us forcibly, in looking
over the second and third reports, that no mention
is made of the works on political economy, commerce, statistics, and the like, that in different
parts of Europe, have, within these few years, increased the mass of knowledge on all these subjects.   Works on legislation are mentioned ; but
no enumeration is subjoined.   We do not observe
that Malthus's Essay on Population is any where
taken notice of.   All this looks as if there were a
class of subjects, and one too of the highest importance to society, that is at present interdicted in
France.   This is weak policy, and unworthy of a
great monarch.   The subjects prohibited will be
only so much the more forcibly imprinted on the
minds of the people.   They will be like the statues
which the jealousy of a Roman emperor excluded
from a procession in which they had a right to appear—" Præfulgebant Cassius atque Brutus,—eo
ipso quod effigies eorum non visebantur."

But whatever be the case with this branch of
knowledge, it is but fair to state, that the physical

and mathematical sciences, and many parts of literature, have been cultivated in France and in the rest of Europe, to great effect, during the last nineteen years, notwithstanding the agitation and distress which have every where prevailed. We are certainly not of the number to whom the Emperor alludes, who pretend that science is retrograde, because they wish it to be extinguished. We rejoice to think that it cannot be extinguished ; and that this is a revolution which no individual is sufficiently powerful to effect. Indeed, we have reason to think, that those branches of knowledge that are least favoured by the Emperor, and to which his protection is not extended, are at this moment studied in France with great assiduity.

FINIS.

REVIEW

LAMBTON'S MEASUREMENT

OF AN ARCH OF THE MERIDIAN.

REVIEW

OF

LAMBTON'S INDIAN SURVEY. *

———

THE measurement of the distance between the meridians of Paris and of Greenwich in 1787, formed
a new era in the art of Trigonometrical Surveying.
The instruments employed in that operation were
of such a superior construction, as to afford a measure of many quantities which were before only
known from theory to exist. Though it was perfectly understood that the three angles of a triangle on
the surface of a spherical body like the earth, must
necessarily exceed two right angles, yet a quantity
so minute as to bear the same proportion to four
right angles which the area of the triangle bore to
half the superficies of the globe, had eluded the
best instruments yet applied to the purposes of
practical geometry. It was not till the survey just
mentioned, that the new theodolite of Ramsden, in

———

* From the Edinburgh Review, Vol. XXI. (1813.)—ED.

the hands of General Roy, and the repeating circle
of Borda, in those of the French mathematicians,
were able to measure a quantity, where even fractions of a second must be accurately ascertained.
The exquisite division of the former of these instruments, and the power possessed by the latter,
of not only measuring any angle, but any multiple
of it, and any number of multiples, rendered them
perfectly equal to such delicate observations.   The
advantage of this was quickly perceived; for the
spherical excess, or the excess of the three angles
of the triangle above two right angles, depending
entirely on the area of the triangle, could be estimated with sufficient accuracy before the angles
were correctly determined, and therefore might
serve for a check on the observations, as effectual
as that which is furnished by the well known property of plane triangles, that the three angles are
always equal to 180 degrees. This was remarked by
General Roy, and applied to the purpose of estimating the accuracy and correcting the errors of
his observations.   The French geometers carried
their views farther; and in seeking to turn the
knowledge of this limit to the greatest advantage,
Legendre discovered, that if each of the angles of
a small spherical triangle be diminished by one-
third of the spherical excess, their sines become
proportional to the lengths of the opposite sides of

8

the triangle, so that the ratios of the sides may be found by the rules of plane trigonometry.

In a science where all the parts are necessarily connected with one another, one improvement can seldom fail of leading to many more. It now became evident, that to carry, through the whole process of a trigonometrical survey, the same accuracy that was employed in measuring the angles of the triangles, methods of calculation must be introduced to which it was before quite unnecessary to resort. Thus, if the object was the measurement of an arch of the meridian, the reduction of the sides of the triangles to the direction of that line, by the usual method of letting fall perpendiculars on it, from the extremities of those sides, and finding the lengths of the parts intercepted, by the rules of plane trigonometry, did not possess a degree of accuracy equal to that which belonged to other parts of the process. The perpendiculars drawn to the meridian from any two points are not in strictness to be regarded as straight lines, but as arches of two great circles perpendicular to it, which would meet if produced in the pole of the meridian, or in the point of the horizon which is due east or west from the place of observation. It is therefore by the solution of a spherical triangle, of which the sides are nearly quadrants and the base very small, that the reduction required is to be made. This is the method followed by Delambre in the measure-

B b

ment of the great arch of the meridian carried across France, for the purpose of determining the length of the *mètre*.   It is a refinement which was not thought of by General Roy ; and we are not sure that it has been followed by any of the geometers who succeeded him in the conduct of the British survey.   It is one however which, when the utmost accuracy is aimed at, ought not to be neglected, especially in high latitudes, where the convergency of the meridians is considerable.

Another refinement, which one should suppose might be even more easily dispensed with than the former, applies to the measurement of the base from which the sides of the triangles are determined.   That line is usually measured by placing rods of equal lengths, or chains stretched with great care, at the ends of one another, for a distance of five or six miles.   It has been usual to consider the base, thus measured, as a straight line, the length of which is just equal to the sum of the lengths of all the rods or chains which have been consecutively placed at the ends of one another. The truth however is, that these rods have not been placed exactly in the same straight line, and that they constitute the sides of a polygon inscribed in a circle, the radius of which is the radius of curvature of the earth at the point, and in the direction in which the base is extended.   The line measured is therefore, in fact, an arch, passing through the angles of this polygon ; and this arch, which is the real base, is longer than the sum of the rods or chains. It is, however, easy to see that the deduction of the real length from the apparent, is not, in this case, a matter of much difficulty.

There is another way of including these corrections, which has been thought preferable by some geometers, and is recommended by the authority of Delambre. According to it, the spherical angles, or those actually measured, are reduced to the angles of the chords; and thus the lengths of the chords are calculated by plane trigonometry, and thence the lengths of the arches themselves are afterwards deduced. The base, measured as above, is also reduced on the chord. This method, though less direct than the former, has considerable advantages in calculation. It was followed by Major Lambton in the survey of which we are now to treat.

A third source of inaccuracy, which had never before been thought of, drew the attention both of the French and English mathematicians engaged in the survey. Triangles, as we have seen, on the surface of the earth, cannot be regarded as plane triangles, because the plummets at the three angular points are not parallel to one another, and of course, the theodolites at these three stations can neither be in the same, nor in parallel planes. But neither can they always be regarded as spherical triangles ; for the plummets at the three angles of the triangle do not all tend to the same point in the interior of the earth ; and, in some cases, do not any one of them intersect another. Spheroidal triangles must therefore differ from spherical ; and though, in such triangles as usually occur in a trigonometrical survey, the difference is of no account, yet there is one case where it can by no means be neglected. This happens, when the bearings of any obtuse line, or rather arch, with respect to the meridians that pass through its extremities, are known, and also the latitude of one of those extremities, and it is required to find the difference of the longitude of the said extremities, or the angle which the meridians passing through them make with one another at the pole. If the base of this triangle is considerable, and very oblique to the meridians, the directions of gravity at its extremities will not intersect the earth's axis in the same point,—and the difference may be so great that it cannot be neglected in calculation.

These corrections have all been taken into account, and the application of them fully exemplified, in the measurement of the great arch between Dunkirk and Formentera, (the southernmost of the Balearic Isles.) Indeed, the book in which the facts and investigations respecting this measurement have been recorded by the French mathematicians, the Base Métrique, is one of the most valuable works which has yet distinguished the beginning of the nineteenth century.

Major Lambton, who, in 1801, proposed the survey of the Peninsula of India, was fully aware of all those new improvements, and perfectly prepared for carrying them into effect. It is indeed much to the credit of the British army, that in a detachment of it, in a distant country, an officer should be found already prepared for a service implying such scientific acquirements, as nothing but the strong impulse of genius could have rendered compatible with the duties or the amusements of a military life.

The plan having been first approved by the governor of Madras, and afterwards communicated to the Asiatic Society, was published in the 7th volume of their Researches. The recent conquest of the Mysore had just opened the interior of the country, and made it practicable to join the coasts of Malabar and Coromandel, by a series of triangles which might be extended on the south, to the extremity of the Peninsula, and to an indefinite distance on the north. It was proposed to execute the work on a plan similar to that pursued in France and England, paying attention to the spherical excess, the spheroidal figure of the earth, and the other circumstances which have just been mentioned. The India Company furnished the Major with the best instruments that could be procured; and indeed it is but justice to remark, that, in whatever concerns geographical improvement, the liberal and enlarged views of the present rulers of India cannot be too highly commended. *   At the present moment, no country in the world, except France and England, has its geography ascertained by a survey so accurate and extensive as that of which we are here to give an account.

The instruments used in the Indian survey, are of the same kind with those employed in the British. The theodolite is one made by Cary, after the model of that invented by Ramsden, and described by General Roy in the Philosophical Transactions for 1790, with some additional improvements. The instrument for the celestial observa-

---

* They have sent out parties, in all directions, for the purpose of ascertaining the bearings and distances of the places which compose or limit their extensive dominions. A late volume of the Asiatic Researches contains an account of the march of an officer, at the head of a detachment, into one of the most remote and unknown districts of India, for no other purpose but to decide a question, interesting only to philosophers, viz. Whether the Ganges rises within or without,—that is, on the south or the north side of the great chain of Himmaleh, the Snowy Mountains, or the Immaus of the ancients?—There are but few of the most enlightened cabinets in Europe which can boast of an expedition equally disinterested and meritorious.

tions, was a zenith sector of five feet radius by the
same artist ; it is capable of ascertaining small
fractions of a second, and appears to be an excellent instrument, though not so large as that used
in the British survey. The chains employed in
the measurement of the bases, were also similar to
Ramsden's. That every source of error might, as
far as possible, be removed, the angles were usually taken three or four times ;—at each time the
angle was read off from the opposite microscopes of
the theodolite, and the results set down in two separate field-books. The mean of the numbers
from the two books, are those employed in the calculation, and recorded in the printed table of observations.

In a survey of the kind here proposed, four separate processes, different in themselves, and
directed to distinct objects, are necessary to be
combined. The first is the measurement of a base,
or of more bases than one, each of which must be
a straight and level line, at least five or six miles
long. This, it has been usual to measure, by
placing straight rods, sometimes of deal, sometimes
of metal, or even of glass, all of the same length,
one at the end of another, each supported horizontally along the whole line. It was found by
General Roy, that a steel chain, made in a particular manner, somewhat like a watch chain, and
stretched in a wooden trough, by weights that are

always the same, is not less to be depended on than
the rods, and is far more convenient. This method
of measuring the base was employed by Major
Lambton, and he considered the base he had
measured, conformably to what is before mentioned, as a polygon or a series of chords inscribed in a
circle, as many in number as there have been
chains. The real base is the circular arch in which
these chords are inscribed.

The next part of the process is the formation of
a series of triangles which go from hill to hill, over
the whole space to be included in the survey, and
having the base already measured, for a side of one
of them. In each triangle the angles are to be
taken, and then, by trigonometry, their sides can
be determined : The whole may be laid down on
paper ; and the position of every point within the
survey may be found, with respect to every other.
This is sufficient, therefore, for determining the
position and magnitude of every line, and every
figure, within a given extent ; but it does not determine the position of the tract surveyed, in respect of the other parts of the earth's surface.   It
does not determine its situation in respect of the
quarters of the heavens, in respect of the parallels
of latitude, or in respect of the different meridians
which divide the surface of the globe from north
to south.   The first of these objects is obtained by
observing carefully the azimuths of one or more of

the sides of the triangles, that is, their bearings with respect to the meridian. This serves to place the whole in its due direction with respect to the cardinal points, or to *orient* the plan, if we may borrow a term from the French, which we wish we had weight enough to introduce into our own language. *

The next thing to be done, is to place the tract surveyed, between the same two parallels of latitude on the artificial globe, which it actually lies between on the surface of the earth. This is done, by observing the latitude at any two stations in the survey, at a considerable distance north or south from one another. If, when this is performed, the distance between the two places reduced to the direc-

---

* We want very much a verb to denote the act of determining the position of a line, or a system of lines, in respect of the quarters of the heavens. The French use the word *orienter* for this purpose; and we propose to translate this by the phrase *to orient*. The English language is remarkably poor in words denoting position in respect of the heavens. Our sailors have been obliged to borrow the harsh term, *rhumb*, from the Portuguese; to denote, by a single word, the point of the compass on which a ship sails. In Scotland they use the word *airth* or *airt*, for the same purpose; and sometimes convert it into a verb, to *airt, orienter*, or *to orient*. The Scots term, however, is neither of so good a sound, or so classical an origin, as that which we propose to introduce.

tion of the meridian be computed, we have the measure of a degree; which, therefore, is a thing almost necessarily implied in a trigonometrical survey.

The position of the whole then, as to its distance from the equator, or from the pole, is thus found; but its distance east or west, from some given meridian, that is to say, its longitude, remains to be determined; and this must be settled by the comparison of the time in some point within the survey, with the time as reckoned under the given meridian. To all these objects Major Lambton has directed his observations, and, we think, with remarkable success.

The base was measured on a plain near Madras, at no great distance from the shore, and nearly on the level of the sea, in spring 1802. The length of the base, reduced to the level of the sea, and to the temperature of 62°, is 40006.44 feet, or 7.546 miles; the latitude of the north end was 13° 0′ 29″, (*Asiatic Researches*, Vol. VIII. p. 149, &c.;) and it made an angle of little more than 12′ with the meridian. From this a series of triangles was carried, about 85 miles eastward, north as far as the parallel of 13° 19′ 49″ N., and south to Cuddalore, latitude 11° 44′ 53″, embracing an extent of about 3700 square miles. The triangles seem well contrived for avoiding very acute and very obtuse angles; the sides of many are from 30 to 40 miles in length, which indicates a fine climate, where the air is very transparent, and a country where hills of considerable elevation are easy to be found.   In computing the sides, Major Lambton reduced the observed angles to the angles of the chords, according to the method of Delambre; and though he computed the spherical excess, he did not use it in any other way than as a measure of the accuracy of his observations.   The knowledge of this spherical excess enables one, from having two angles of a spherical triangle, such as occurs in the survey, to find the third, though it be not observed.   This is a facility of which a careful observer will avail himself as seldom as possible, as it deprives him of the check by which the errors in the angles might be detected.   The difficulty of the country often proves a temptation to make use of it in this way, so as to avoid the necessity of carrying the theodolite to the more inaccessible points. Major Lambton has no appearance of a person who would save labour at the expence of accuracy; and, whenever he has omitted to take all the three angles of a triangle, we believe that it has arisen from the necessity of the case.   The chords, which were the sides of the triangles, were then converted into arches; and as by a very judicious arrangement, which, however, is not always practicable, Major Lambton had contrived, that the sides of the four triangles which connected the stations at the

south and north extremities should lie very nearly in the direction of the meridian, their sum, with
very little reduction, gave the length of the intercepted arch, which was thus found to be 95721.326
fathoms.

By a series of observations for the latitude, at
the extremities of this arch made with the zenith
sector above mentioned, the amplitude of the arch
was found to be 1°.58233, by which, dividing the
length of the arch just mentioned, Major Lambton
obtained 60494 fathoms for the degree of the meridian, bisected by the parallel of 12° 32′. This,
till the survey was extended farther to the south,
was the degree nearest to the equator, (except that
in Peru, almost under it,) which had yet been measured, and was, on that account, extremely interesting.

The next object was to measure a degree perpendicular to the meridian, in the same latitude.
This degree was accordingly derived from a distance of more than 55 miles, between the stations
at Carangooly and Carnatighur, nearly due east
and west of one another. Very accurate measures of the angles, which that line made with the
meridian at its extremities, were here required;
and these were obtained, by observations of the
pole star, when at its greatest distance from the
meridian. For this purpose, a lamp was lighted,
or the blue lights were fired at a given station, the

azimuth of which was found by the pole star observations, and afterwards its bearing in respect of
the line in question. Thus the angle which the
meridian of Carangooly makes at the pole, with
that of Carnatighur, or the difference of longitude
of these two places, was computed. It was then
easy to calculate the amplitude of the arch between them ; and thence the degree perpendicular
to the meridian at Carangooly, was found to be
61061 fathoms.

With regard to the measure of this perpendicular degree, we confess that we do not see reason to
place great confidence in it, notwithstanding our
high opinion of the observer. The method of determining the difference of longitude, by the convergency of the meridians, or the angles they make
with a line intersecting them, is not easily applicable in low latitudes, or in places near to the equator ; because there, a very small error in the observation of the azimuths, must produce a very
great one in the difference of longitude. The convergency of the meridians is so small, in the present instance, that if a line were to be drawn
through Carangooly parallel to the meridian of
Carnatighur, it would not make with the former
an angle of one minute. A very small error,
therefore, in ascertaining the angle which these
lines make with a third line, must greatly affect
the quantity of the angle which they make with

one another, This is also evident from considering, that at the equator, all the meridians make right angles with the line from east to west, and have therefore no convergency at all. The problem of determining the difference of the longitude, or the arch of the equator, by the angles which it makes with the meridian, comes here under the *porismatic* or indeterminate case, where the data can lead to no definite conclusion. This is evidently true at the equator ; and we are con. stantly coming nearer to this condition of things, as we come nearer to that circle. The porismatic case of a problem, like every other, does not arise all at once, but comes on by gradations ; every approach to the state in which the thing sought is quite indeterminate, being marked by the greater looseness and inaccuracy of the determination actually given.

Of the degree of the perpendicular as here given, viz. 61061 fathoms, we have farther to remark, that when compared with the degree of the meridian, it brings out the compression at the poles equal to $\frac{1}{210}$, which is certainly much too great. But if it be diminished by 200 fathoms, and reduced to 60861, as an ingenious writer *(Phil. Trans.* 1812, p. 342) contends that it ought to be, on account of an error in calculation, which has escaped Major Lambton, it gives for the compression $\frac{1}{530}$, which is probably not far from the truth.

The measures of which we have been giving an account, were made in 1803; the next, of which we are informed in the tenth volume, were in 1806, when the series of triangles was carried quite across the Peninsula to the Malabar coast, which they intersected at Mangalore on the north, and Tillicherry on the south. In this tract they of course passed over the Ghauts, so remarkable both in the natural and civil history of Hindostan; and as the stations, most probably, are the tops of some of the highest mountains, their heights may serve to give some idea of the general elevation of the chain. The most considerable are, Soobramanee and Taddiandamole, in the western Ghaut, not very far from the coast, the former 5583 feet, and the latter 5682 above the level of the sea. Considerable difficulty could not fail to be experienced in conducting the survey across these mountains.

" I had laid (says the Major) the foundation for a southern series of triangles, to be carried through the Koorg, to Mount Delli, (on the coast,) which was rendered practicable by the assistance afforded me by the Koorg Rajah, to whose liberal aid I am indebted for the successful means I had in carrying the triangles over those stupendous mountains." Vol. X. p. 295.

The heights of the stations were all determined

from the distances and observed angles of elevation; and it is no small proof of accuracy, that
after ascending the chain of the Ghauts, from the
Coromandel coast on the east, and descending from
it to the level of the sea on the Malabar coast, a
distance in all of more than 360 English miles,
they found the sum of all the ascents, and of all the
descents, reckoned from the level of the sea, to
differ from one another only by eight feet and a
half.   This is the more remarkable, that the
angles of elevation and depression, on account of
the refraction, are the parts of trigonometrical
measurement, in which error is most difficult to be
avoided.   In every case the angles of elevation and
depression between the same objects were constantly measured; and thence the refraction was
determined; the double of it being equal to the
apparent elevation, *plus* the horizontal distance in
minutes, *minus* the apparent depression.   The re-
fraction seems to have varied from $\frac{1}{4}$ to $\frac{1}{20}$ of the
horizontal arc; but as the heights of the barome-
ter and thermometer, at the time the angles were
measured, are not put down, no inference can
be drawn as to the relation between the density
of the air and the quantity of the terrestrial refraction.

From the triangles thus carried across the Peninsula, a correct measure of its breadth was obtained,
and one considerably different from what was be-

fore supposed. The distance from Madras to the opposite coast, in the same parallel, is 360 miles very nearly ; the best maps, till then, made it exceed 400.

It now became proper to measure a second base in the interior of the country, which was accordingly done near Bangalore, about 170 miles west from Madras, not far from which the first base was measured. The execution of the work was committed to Lieutenant Warren of the 33d Regiment ; and no better proof of the accuracy of the whole combined operation is necessary, than that when the length of this base was deduced from the base at Madras, 170 miles distant, by means of the intervening triangles, the computation exceeded the actual measurement only by 3.7 inches. The length of this base, reduced to the level of the sea, is 39793.7 fathoms $= 7.536$ miles. The same precautions were employed here as in the preceding measurement, and even with increased attention. For ascertaining the latitudes with the zenith sector, a number of different stars were observed; the same stars were observed, at both stations, a great many different times, alternately, with the face of the instrument toward the east and toward the west ; so that the error of the line of collimation was completely destroyed. The observations, with the face of the instrument turned the same way, are usually very near to one another ; so that an error of a second in the determination of the latitude, can hardly be supposed. From these observations, the degree of the meridian came out 60498 fathoms, in lat. 12° 55' 10".

The next thing attempted, was the determination of a degree perpendicular to the meridian in the above latitude, which is that of Savendroog, not far from Bangalore. Here, again, though the operation is conducted with all possible care, and in circumstances that Major Lambton thought more favourable to accuracy than the former measurement of the perpendicular, a similar uncertainty takes place as to the result. The degree perpendicular to the meridian at the place just named, was found to be 60747.8 fathoms; on which Major Lambton remarks, that taking the ratio of the earth's diameters to be 1 to 1.003125, and the meridional degree, in lat. 12° 55' 10", to be 60498 fathoms, the degree of the perpendicular will come out 60858 fathoms, which exceeds the measured degree by 110 fathoms; so that it may be inferred, either that the earth is not an ellipsoid, or that this measure is incorrect. We already stated our objections to the degree of a perpendicular arch, ascertained by the convergency of meridians in such low latitudes. The Major himself seems to think that no great reliance can be had on results so obtained. "The great nicety," he says, "in making the Pole star observations (for the azimuths) is well understood; and it will be made more mani-

fest in the case before us, by increasing or diminishing the half sum of the azimuths reciprocally taken
at Mullapunnabetta and Savendroog, by one second only, when it will appear that a difference of
nearly one hundred and fifty fathoms will be thereby occasioned in the perpendicular degree."

But if the method of measuring a degree of the
perpendicular by the convergency of the meridians
cannot be successfully practised, what method must
be had recourse to ? The measurement of this arch
is very necessary for determining the difference of
longitude, and is therefore an important element in
the survey.   Other methods of ascertaining the
longitude ought no doubt to be tried, such as that
which Major Lambton mentions as having been
strongly recommended to him by the late Astronomer-Royal, by carrying a good time-keeper between
two meridians at a known distance, which I mean,
says he, to put in practice in the course of my future operations.   " I had also," adds he, " de-
vised another method, by the instantaneous extinction of the large blue lights used at Savendroog,
the times of which were to be noticed by observers
at Mullapunnabetta and Yerracondah, the distance of whose meridians on a parallel of latitude
passing through Savendroog is nearly 135 miles.
The experiments were attempted ; but the weather
was so dull that the lights could scarcely be distinguished.   There is, besides, a difficulty in fixing

the precise moment of extinction ; and even in the most favourable state of the atmosphere, when the lights may be distinctly seen with the naked eye, at near seventy miles distance, to come within half a second of the truth, would be as near as the eye is capable of, which is $7''\frac{1}{2}$ in an angle at the pole : But the mean of a great number of observations might come very near the truth."

The Major then proposes the comparison of celestial observations, such as occultations of the fixed stars by the Moon, eclipses of the satellites of Jupiter, &c. for the longitude, to be made at Madras and Mangalore—almost five degrees of longitude removed from one another, and of which the distance in fathoms is perfectly determined from the survey. He concludes with a passage, full of the modesty characteristic of real talent, and breathing the spirit of ardent and persevering research, which nothing but the love of truth is able to inspire.

" In short, the difficulty of obtaining this desideratum, (the knowledge of the true figure of the earth,) and the important advantages to geography and physical science which must accrue from it, are such powerful incitements to the prosecution of the inquiry, that I may venture an assurance of leaving nothing undone, which may come within the compass of my abilities, to give every possible satisfaction on the subject ; and if my endeavours shall prove successful to throw some light on the path of future discovery, I shall close my labours with the grateful reflection, that while employed in conducting a work of national utility, I shall have contributed my humble mite to the stock of general science." P. 368.

The 12th volume of the Asiatic Researches contains an account of the extension of this survey, to the southern extremity of the Peninsula, and the measurement of another considerable portion of the meridian, amounting in all to nearly six degrees; —the longest arch, excepting that in France, which has yet been measured on the surface of the earth. When the work was first undertaken, the principal object was to connect the two coasts of Coromandel and Malabar, and to determine the latitudes and longitudes of the principal places, both on the coasts and in the interior. As the work proceeded, the design was enlarged; and, in addition to the triangles carried across the Peninsula between the latitudes of 12 and 14 degrees, as already mentioned, another series was extended from Tranquebar and Negapatam on the Coromandel coast, across to Paniany and Calicut on the opposite shore; and to render the skeleton complete, a meridional series was carried down the middle of the Peninsula as far as Cape Comorin, from which were extended other series to the east and west along the sea coasts; so that a web of triangles has been completely woven over the Peninsula of India from the paral-

lel of 14° to its utmost extremity. It is to the meridional arch, of nearly six degrees, thence deduced, that this last memoir relates. It was presented
to the President of the Asiatic Society by the Governor General, Lord Minto, who added this judicious and merited encomium: " I have great pleasure in being the channel of communicating to the
learned Society, a paper containing matter of such
high importance to the interests of science, and
furnishing so many new proofs of the eminent endowments and indefatigable exertions which have
long distinguished the character and labours of its
respectable and meritorious author."

In this measurement, the meridian of the Dodagoontah station, or of Savendroog, was continued
south to Punnae, in the latitude of 8° 10′; and the
series of triangles, for the purpose of ascertaining
its length, was continued to the same point. In
the extent of this prolongation, two new bases were
measured, one at Putchapoliam, where the meridian
intersects the parallel of 11°, and another at Tin-
nevelly, near the southern extremity of the arch.
These bases were nearly of the same length, (somewhat shorter than that at Bangalore,) and measured with the same commendable attention to every
circumstance which could ensure their accuracy.
The triangles were carried on in the same manner,
being a part of that great system which we have
already mentioned as covering all this part of the

Peninsula. In many places the country is high and difficult to penetrate; the highest mountain in the whole survey occurs here, viz. the *Hill* (for so it is called) of Permaul in latitude 10° 18′; its height is set down at 7367 feet. A very laudable precaution was taken throughout by Major Lambton, that of describing the positions of the great stations, and giving marks, by which an astronomical instrument may be placed in the same situation with his, if any of the observations should seem to require repetition or verification.

The observations for the latitude appear to have been conducted even with increased diligence. The practice of reversing the sector is never omitted; the latitude of Putchapoliam, the northern extremity of this prolonged meridian, is determined from the mean of 173 zenith distances of stars, all passing very near the vertex. The number of similar observations at Punnae was 226, and from these was deduced the amplitude of the arch between the stations just named, viz. 2° 50′ 10″.5, the length being 171516.75 fathoms. The differences between the zenith distances of the same star seldom exceed 3″, and are usually much less; so that, taking into account the number of observations, it cannot be doubted that these latitudes are determined to a fraction of a second.

On this meridian, the distances of five stations, with the corresponding latitudes, were determined

in the course of the present and the former survey;
Punnae the south extremity ; Putchapoliam, Dodagoontah, Bomasundrum and Paughur, the north-
ernmost point.   The amplitude of the whole arch
was 5° 56′ 47″.32, and its length 359595.4 fathoms.
From this and the other points named above, the
following degrees are deduced :—

		Deg. in Fath.	Mid. Lat.
Punnae and Putchapoliam	-	60473	9.34.44
Punnae and Dodagoontah	-	60496	10.34.49
Punnae and Bomasundrum	{	60462	11. 4.44 }
Punnae and Paughur	- {	60469	11. 8. 3 }
Mean of the two last	-	60465.5	11. 6.23.5
From a former measurement		60494	12.32.0

In these degrees we perceive the same anomalies
which have been observed in France and in England, and which will probably always occur, where
contiguous parts of the same arch are compared
with one another.   The degree in the parallel of
11° 6′ 23″ is 60465.5 fathoms, which is less than
that in the parallel of 9° 34′, a degree and a half
farther to the south.   This is similar to what appears in the degree in England ; and there is an
instance of the same species of retrogradation, when
the parts of the arch between Dunkirk and Formentera are compared with one another.   Some
part of this irregularity, but certainly a very small
one, may be ascribed to error of observation ; the

greatest part must, we think, be placed to account of
the irregularities in the direction of gravity, arising
from the inequalities at the surface, or in the interior of the earth ; the attraction of mountains,
for example, or the local variations of density in the
parts immediately under the surface.   On the effect of these last, Major Lambton remarks, " That
between Dodagoontah and Bomasundrum, (13° and
14°,) there is a vein of iron ore which might be
supposed to have affected the plummet." A more
particular description, however, of the country
would be necessary to enable us to judge of the
probability of this hypothesis.   A mere vein, in the
strict sense of the word, would be a cause inadequate to such an effect as is here ascribed to it ;
but a great *mass* of iron ore, or a body of ferrugin-
ous strata, might be sufficient to produce the effect.
We long ago remarked, in speaking of the trigonometrical survey of England, that it would have
been of great importance to have added to it a mineralogical survey, as the results of the latter might
have thrown some light on the anomalies of the former.   The same thing is suggested by the objects
now under consideration.   It would be extremely
desirable also to have a vertical section in the direction of the meridian and of the perpendicular, at
those places where observations for the latitude are
made.   This might afford a satisfactory solution of
many difficulties which at present are sufficiently

perplexing, and seem to increase just in proportion
to the extent and accuracy of the observations.
Major Lambton goes on to remark, " That the arc
between Putchapoliam and Dodagoontah gave the
length of the degree in latitude 11° 59′ 54″, equal
to 60529 fathoms, while the arch between Putchapoliam and Bomasundrum gave the same degree
only 60449. Both these stations are sufficiently
remote from mountains, to remove all suspicion of
a disturbance from that cause ; but as no doubt remained as to the existence of some disturbing cause,
I attributed it to the effects of the bed of ore, and
concluded that the plummet had been drawn to
the northward at Dodagoontah, and to the south-
ward at Bomasundrum, which would give the celestial arc between Putchapoliam (to the south of
both) and Dodagoontah too little, and that between
Putchapoliam and Bomasundrum too great ; making, of consequence, the length of a degree too
great in the first case, and too little in the second.
Being," he adds, " confident as to the accuracy of
the observations at both places, in consequence of
the circumstances just mentioned, I thought it reasonable to take the mean of the two degrees, which
gave 60490 fathoms for the degree in latitude
11° 59′ 54″."

In the conclusion of the paper, the Major reduces the degrees into a consistent form, and apparently cleared of all irregularity, (p. 94,) but on a

principle of which we cannot entirely approve, as it involves in it too much theory. The mathematical reasoning is correct ; but the introduction of a degree measured in another latitude, though it is quite legitimate in a general inquiry into the figure of the earth, prevents the results of the Indian measurement from appearing as independent facts, resting on the foundation of experiment alone.

The simplest and most unexceptionable way of deducing from a large arch, (the parts of which, as actually measured, are not perfectly consistent,) the results that may be accounted the nearest approximation to the truth, is to consider, that if the elliptic hypothesis be true, whatever be the compression, the successive degrees of the meridian must increase, on receding from the equator, by a quantity proportional to the sine of the double latitude.   Thus, if x be the degree in the latitude L, the next degree is $x + n$ sin. 2L ; the next to that is $x + n$ sin.2L $+ n$ sin. (2L $+ 2°$), &c. where n is a constant quantity, to be determined without the assistance of theory, by assuming different values for it, and adopting that which agrees nearest with the observations. This is easy, because n sin. 2L is always a small quantity.   In the southernmost point of Major Lambton's arch, it is between $2\frac{1}{2}$ and $3\frac{1}{2}$ fathoms : the value that seems to us to answer best, is 3.1 fathoms ; and in this way we deduce the first degree of the arch, that which begins at Punnae, in lat. 8° 39′ 38″, and has its middle in 9° 9′ 38″, equal to 60473 fathoms. This is derived from a comparison of the arch between Punnae and Putchapoliam, which consists of 2° 50′ 10″, and is certainly, as far as observation can go, very accurately determined. In this way, the successive degrees are as follows :

Mid. Lat.	Length.	Mid. Lat.	Length.
9° 10′	60473 fath.	12° 10′	60483.2 fath.
10° 10′	60476.1	13° 10′	60487.2
11° 10′	60479.5	14° 10′	60491.3

These are a little different from Major Lambton's results, to which they would have been brought nearer, if we had employed the arch between Punnae and Dodagoontah, in the determination of the first degree. But as the latitude of Dodagoontah is in all probability affected by the attraction of the plummet toward the north, so that its zenith is carried too far to the south, the arch between it and Punnae must be too small; and therefore we thought it best to avoid this arch in the fundamental determination.*

* To deduce the mean degree from a large arch, such as one of nearly three degrees, by dividing the length of the arch by its amplitude or number of degrees, is not exact, as the degrees increase each above the preceding by the quantity n sin.($2L$ + 2°.) The length of the arch

The anomalies which have occurred in the measures of degrees, and of which the appearances seem to increase in proportion as greater pains are taken to avoid inaccuracy, have naturally drawn the attention of mathematicians ; and the question, what part of them is to be ascribed to error, and what to irregularities in the structure of the globe, has come, of course, to be considered.   That a small part of them only can be ascribed to the former cause, is rendered probable by the very circumstance just stated ; that they are not diminished, nay, that they even seem to be increased, by the care taken to avoid error.   It seems clear, from that consideration, that the irregularities are in the object sought for, and are only brought more in sight by more microscopical observation, by the excellence of the instruments, the accuracy of the computations, and the extent of the lines measured. No measurement was ever executed with greater care than that in France ; and the great extent of the arch measured, as well as the ability and skill of the observers, make the mean result, the length of the degree in the parallel of 45°, the *datum* most perfectly ascertained of any that regards the figure of the earth.   Yet even here, we find in the

---

ought to be diminished by the sum of all these quantities before it is divided by the amplitude ; and this division gives not the degree in the middle of the arch, but that at the beginning of it, or the farthest to the south.

detail that there are great anomalies, and that the
successive degrees increase with much irregularity.

The arch between Greenwich and Dunkirk gives
the degree greater than that which is derived from
the arch between Dunkirk and the Pantheon at
Paris by 7.23 toises; the next difference is 8.4;
then 32.4, 12.9; and lastly—2 from the arch between Montjouy and Formentera. In this last
case, there is an absolute retrogradation; and the
degree increases on going to the south, just as it is
observed to do in the arch measured in England,
and in that measured in Hindostan.

The irregularities in the French measurement
induced Delambre to scrutinize the latitudes of all
the above places with the utmost care; but he
could find nothing sufficient to account for the irregularities. (See Base Métrique, Tom. III. p. 84.)
The observation of the latitude at Montjouy appeared exact; yet, when compared with one at
Barcelona, very near to Montjouy, an error of
3″.24 was discovered; and Delambre, apparently
with much reason, considers this difference as a certain proof of the irregularities of the earth. To
the same cause he ascribes the rest; and, indeed,
from the very progress which they hold, some local
affection seems necessarily suggested.

The consequence of all this is, that for the
whole of the arch in France, the degrees are best
represented by supposing a compression of $\frac{1}{150}$, or

8

$\frac{1}{148}$; while, by taking in a greater range, and comparing the degrees in France with those in distant countries, the compression comes out less than the half of this, viz. $\frac{1}{320}$, or $\frac{1}{309}$. To reconcile the measures actually made with a compression of $\frac{1}{320}$, it will be necessary to make the following corrections on the latitudes :—For Paris, 0 ; Montjouy, + 3″.6 ; Carcassonne, + 0.88 ; Dunkirk, + 3″.06 ; and for Evaux, +5″.83. These are wholly improbable as errors of observation, and must be attributed to local attractions, which act irregularly on the plumb line.—Bâse Mètrique, ib. p. 92.

The same thing may be said of the arc measured in England by Colonel Mudge : the whole arc, taken together, agrees very well with the measures in France, and with that in Lapland, as lately ascertained by the Swedish academy. * But if the parts of this

* We have compared together the five arches of the meridian, which, from their extent, and all other circumstances, seem the best entitled to confidence, viz. that in Peru, by Bouguer and Condamine; in Hindostan, by Major Lambton; in France and England, comprehending the whole extent, from the parallel of Greenwich, to that of Formentera, by Delambre and Mechain, and in part by General Roy; that in England afterwards, by Colonel Mudge; and, lastly, that in Lapland, by M. Swanberg ; and the results which we have found are extremely consistent, and give, for the compression at the poles, $\frac{1}{312.5}$. When this compression is adopted, there does not appear an error of more than 9 fathoms in the measure of any of the above degrees. The

arc be compared, an irregularity is found, and the degrees appear to increase on going from the north to the south. In giving an account of Colonel Mudge's measurement in a former Number of this Journal, we ascribed the fact just mentioned, to local irregularities in the direction of gravity, and we still consider this as by far the most probable supposition. A paper, however, written with great knowledge of the subject, and full of sound mathematical reasoning, has been published by Don Rodriguez in the Philosophical Transactions for 1812, which is quite on the opposite side, and ascribes the irregularities in the arc to errors of observation. Don Rodriguez, if we mistake not, is one of two Spanish gentlemen who accompanied MM. Biot and Arago, and assisted in the operations by which the meridian that had been traced through France was extended to the southernmost of the Balearic Isles. He seems perfectly acquainted with the methods of calculation, and all the most recent improvements which respect the problem of

---

French, from their own measures in France and Peru, bring out a compression of $\frac{1}{309}$ nearly. Thus the results are consistent with the supposition that the earth is an elliptic spheroid, when the arches compared are large and distant from one another : when they are small, and near to one another, they do not agree with that hypothesis, nor indeed with any other single hypothesis that can be laid down. This is what might be expected, and does not invalidate the general conclusion.

4

the figure of the earth.   We do not think that he has proved that the irregularities in this measurement arise from errors of observation ; and we are of opinion, though the amount of these irregularities may now be more exactly estimated than before, that with regard to their cause, the question rests precisely where it did.   But though we are not convinced by Don Rodriguez, we must do him the justice to say, that his argument is fairly conducted, and that he has displayed great knowledge of the subject, and perfect familiarity with the best methods hitherto employed in the solution of this difficult problem.   We have therefore observed with regret, that this ingenious foreigner has been attacked in some of the English Journals, with a violence and asperity which the subject did not call for, and which his paper certainly did not authorize.

When there are unlooked-for results in any system of experiments or observations, the errors into which the observer may have fallen, naturally come to be considered as affording one solution of the difficulty.   We are not to suppose, that any man engaged in experimental investigations, can be exempted from such an inquiry ; nor, when such inquiry is instituted, are we to suppose that he is subjected to a personal attack.   The principle on which Don Rodriguez proceeds, though it may be erroneous, seems to be general ; it is applied equally to the French and the English mathematicians ; and the anomaly of more than 3″ in the latitude of Montjouy, is ascribed by him, not to local irregularity, but to the mistake of Mechain, a man eminently skilled in the art of astronomical observation. The calm and dispassionate memoir of the Spanish mathematician, does not therefore give any ground for supposing it to be meant as a personal attack, and still less as a national one.

We observe, with pleasure, however, that the true resolution of the difficulty is most probably at hand.   The continuation of a meridional arch must afford the best means of discovering from what cause the irregularities observed in it arise. If they arise from physical irregularities in the structure of the globe, or in the direction of gravity, a compensation in the course of a great arch may be expected to take place.   If a body of heavy matter, at any point, make the plummets on each side of it converge more than they ought to do, the zeniths will be carried too far off from one another ; the amplitude of the celestial arch will be increased ; and the length of the terrestrial degree will, of course, be diminished.   But as the zenith on one side of this point was carried too far to the south, and on the opposite too far to the north, the degrees on either side will be rendered too great, the amplitudes of the celestial arches being made too small.   Thus an opposite error will take place, and what is added to one degree will probably be taken from the next. This is not likely to happen if the errors arise from inaccuracy of observation : these errors will not be as any *function* of the distance, but, depending on accident, must be quite irregular in their distribution. It is with pleasure, therefore, that we see a meridian which has been extended from the shores of the British channel along the west side of England, viz. the meridian of Delambre now produced into Scotland, where it falls on the east side of the island, and is about to be continued till it intersect the shores of the Murray Firth, or the Northern Ocean. The combined arches in France and England will then extend nearly to 20 degrees ; and in a few years we shall perhaps see the distance between the parallels of the Balearic and the Orkney Islands, ascertained by actual mensuration. We believe that this important operation could not easily be in better hands than those in which it is actually placed ; and, when it shall be completed, the British army—in General Roy and the officers who have succeeded him in the conduct of the English survey—and in Major Lambton whose works we have been now treating of, will have the glory of doing more for the advancement of general science, than has ever been performed by any other body of military men.

FINIS.

REVIEW

OF

LAPLACE, ESSAI PHILOSOPHIQUE

SUR LES PROBABILITÉS.

REVIEW

OF

LAPLACE, SUR LES PROBABILITES. *

It is to the imperfection of the human mind, and not to any irregularity in the nature of things, that our ideas of chance and probability are to be referred. Events which to one man seem accidental and precarious, to another, who is better informed, or who has more power of generalization, appear to be regular and certain. Contingency and verisimilitude are therefore the offspring of human ignorance, and, with an intellect of the highest order, cannot be supposed to have any existence. In fact, the laws of the material world have the same infallible operation on the minute and the great bodies of the universe ; and the motions of the former are as determinate as those of the latter. There is not a particle of water or of air, of which the condition is not defined by rules

* From the Edinburgh Review, Vol. XXIII. (1814.)—Ed.

as certain as that of the sun or the planets, and
that has not described from the beginning a trajectory determined by mechanical principles, subjected to the law of continuity, and capable of being
mathematically defined. This trajectory is therefore in itself a thing *knowable*, and would be an object of science to a mind informed of all the original conditions, and possessing an analysis that could
follow them through their various combinations.
The same is true of every atom of the material
world ; so that nothing but information sufficiently
extensive, and a calculus sufficiently powerful, is
wanting to reduce all things to certainty, and, from
the condition of the world, at any one instant to deduce its condition at the next ; nay, to integrate
the formula in which those momentary actions are
included, and to express all the phenomena that
ever have happened, or ever will happen, in a
function of duration reckoned from any given instant. This is in truth the nearest approach that
we can make to the idea of Omniscience ; of the
Wisdom which presides over the least as well as the
greatest things ; over the falling of a stone as well
as the revolution of a planet ; and which not only
numbers and names the stars, but even the atoms
that compose them.

The farther, accordingly, that our knowledge
has extended, the more phenomena have been
brought from the dominion of Chance, and placed

under the government of physical causes ; and the
farther off have the boundaries of darkness been
carried.   It was, says Laplace, of the phenomena not supposed to be subjected to the regulation of fixed laws, that superstition took hold, for
the purpose of awakening the fears and enslaving
the minds of men.   The time, adds he, is not far
distant, when unusual rains, or unusual drought,
the appearance of a comet, of an eclipse, of an
aurora borealis, and, in general, of any extraordinary phenomenon, was regarded as a sign of the
anger of heaven ; and prayers were put up to avert
its dangerous consequences.   Men never prayed
to change the course of the sun or of the planets,
as experience would have soon taught them the inefficacy of such supplications.   But those phenomena of
which the order was not clearly perceived, were
thought to be a part of the system of nature which
the Divinity had not subjected to fixed laws, but
had left free, for the purpose of punishing the sins
of the world, and warning men of their danger.
The great comet of 1456 spread terror over all
Europe, at that time alarmed by the rapid succes-
ses of the Turks, and the fall of the Greek empire ;
and the Pope directed public prayers to be said on
account of the appearance of the comet, no less
than the progress of Mahomet.

It is curious to remark how different the sensa-
tions have been which, after four revolutions, this

12

same comet has excited in the world. Halley having recognized its identity with the comets of 1531, 1607, 1682, showed it to be a body revolving round the sun in 75 years nearly ; he foretold its return in 1758, or the beginning of 1759, and the event has verified the most remarkable prediction in science. Comets have since ceased to be regarded as signs of the Divine displeasure ; and every body must have remarked, with satisfaction, how far the comet of 1811 was from being viewed with terror, (in this country at least,) even by the least instructed of the people, and from exciting any sentiment but admiration of its extraordinary beauty. The dominion of Chance is thus suffering constant diminution ; and the *Anarch Old* may still complain, as in Milton, of the encroachments that are continually making on his empire.

Probability and chance are thus ideas relative to human ignorance. The latter means a series of events not regulated by any law that we perceive. Not perceiving the existence of a law, we reason as if there were none, or no principle by which one state of things determines that which is to follow. The axiom, or, as it may be called, the definition, on which the doctrine of probability is founded, is, that when any event may fall out a certain number of ways, all of which, to our apprehension, are equally possible, the probability that the event will happen with certain conditions accompanying it, is expressed by a fraction, of which the numerator is the number of the instances favourable to those conditions, and the denominator the number of possible instances. Thus, the probability of throwing an ace with one die is denoted by $\frac{1}{6}$, as there are six ways that the event may turn out, and only one in which it can be an ace. With two dice, the chance of throwing 2 aces is $\frac{1}{16}$; as each face of the one ace may be combined with any face of the other. Certainty is denoted here by unity; it is what happens when all the cases are favourable to the condition expected, and when the numerator and denominator of the fraction are the same. It were absurd to say, that the sentiment of belief produced by any probability, is proportional to the fraction which expresses that probability; but it is so related, or ought to be so related to it, as to increase when it increases, and to diminish when it diminishes.

The calculation of *Probability* is therefore a very ingenious application of mathematical reasoning, in order to substitute for that certainty which is quite beyond our reach, the degree of evidence that the case admits of, and to reduce this to a system of accurate reasoning. The thing obtained is only probability; but we have a certainty as to the degree in which it exists.

The invention of this calculus does not go far back. It is true, that wherever there have been

games of chance, and they have been in all countries from the rudest to the most civilized, there
must have been some numerical estimate formed of
the probability of certain events, by which the
stakes and the expectations of the gamesters must
have been regulated.   The principle just stated,
must therefore with more or less distinctness have
been long recognized ; but nothing like a system
of reasoning founded on it is to be found before the
time of Fermat and Pascal.   Huygens was the
next after these two illustrious men who treated of
this matter in a treatise, *De Ratiociniis in Ludo
Aleæ*.   Several other mathematicians, Huddes and
De Witt in Holland, Halley in England, applied
the same calculus to the probabilities of human life,
and the latter published the first tables relative to
that object.   James Bernoulli, about the same
time, proposed and resolved many problems concerning probabilities, and composed the treatise entitled *Ars Conjectandi*, which was not published
till 1713, some years after his death.   This work
is worthy of the high reputation of the author, who
treats in it of the probability which a succession of
the same events, at any time, gives of its continuance ; and he was the first to demonstrate a proposition concerning the indefinite multiplication of
casual events, to which we shall again have occasion to advert.   Monmort published an estimable
work on the same subject, *Essai sur les Jeux de*

Hasard ; and Demoivre followed with his treatise *On the Doctrine of Chances*, which first appeared in the Philosophical Transactions for 1711, but was afterwards published in three editions, which the author successively improved. This work is the first that mentioned the theory of *recurring series*, a subject of such importance in algebra, and connected with so many of the discoveries which have since been made in the calculus of *Finite Differences.* Laplace does great justice to it, and has entered into an analysis of the part that relates to series. Demoivre gives a demonstration of the theorem of Bernoulli, just referred to, which, in a series of events, serves to connect the future and the past. Several other mathematicians, and particularly Lagrange, have been attracted by the results which this theory offered, and by the difficulty of the problems it suggested, which seemed in many respects to require a new application of analysis. The last who has treated of it, is our author himself, in a large work in quarto, entitled *Théorie Analytique des Probabilités,* published at Paris in 1812. The essay now under review, is an abstract of this last, containing an account of the more important conclusions deduced in it, together with many general and profound remarks on the principles of the calculus, and their application to the researches of philosophy, as well as to the affairs of life.

The analytical work contains some valuable improvements in this branch of the mathematics. We have adverted to the use made by Demoivre, in his work on Chances, of the series, called Recurring, in which the coefficient of each term is formed in the same manner from the coefficients of a certain number of the preceding terms. The generalization of this property led Laplace to consider all those series in which the coefficients are formed by substituting the exponents, every where, in the same formula ; or where, in every term, the coefficient is the same function of the exponent. A series of this kind being supposed, a function of the variable quantity may be found, from the developement of which the series may be derived ; and this function is what Laplace calls the Generating Function *(Fonction Génératrice)* of the coefficients in the supposed series, or rather of the function in which all those series are included. This gives rise to a new branch of analysis, the calculus of Generating Functions, the principles of which he first explained in the Memoirs of the Academy of Sciences for 1779. From these series, by applying the method of finite and partial differences, he has extracted results that throw great light on the Doctrine of Chances, and readily afford demonstrations of many propositions that cannot but with the greatest difficulty be proved by any other means. It must not seem surprising that the

Doctrine of Series is thus intimately connected with the Theory of Probabilities; for it should be remembered, that the first considerable improvement in that theory came from the same quarter. The numbers of combinations that can be formed of a given number of things, taking them two and two, three and three, &c. are given by the successive coefficients of a binomial raised to the power denoted by the number of things in question. Such combinations are evidently much concerned in the laws of chance; and Bernoulli deduced from them a great number of conclusions concerning those laws. Demoivre went farther than Bernoulli, and Laplace much farther than either; but to give any adequate idea of the analytical methods which he has employed, is not to be expected in an abstract like the present. For a general view of the analytical methods applied to the calculation of probabilities, we may refer the reader to the conclusion of the Essai Philosophique, p. 90, &c., and to the beginning of the Théorie Analytique. To a passage in the latter, however, we cannot but advert, and with much less satisfaction than we have generally felt in pointing out any of the remarks of this celebrated writer to the attention of our readers. " *Il paraît que Fermat, le véritable inventeur du calcul différentiel, a considéré ce calcul comme une dérivation de celui des différences finies,*" &c. Against the affirmation that

Fermat is the real inventor of the differential calculus, we must enter a strong and solemn protesta-
tion. The age in which that discovery was made,
has been unanimous in ascribing the honour of it
either to Newton or Leibnitz; or, as seems to us
much the fairest and most probable opinion, to
both; that is, to each independently of the other,
the priority in respect of time being somewhat on
the side of the English mathematician. The writers of the history of the mathematical sciences have
given their suffrages to the same effect;—Montu-
cla, for instance, who has treated the subject with
great impartiality, and Bossut, with no prejudices
certainly in favour of the English philosopher. In
the great controversy, to which this invention gave
rise, all the claims were likely to be well considered; and the ultimate and fair decision, in which
all sides seem to have acquiesced, is that which has
just been mentioned. It ought to be on good
grounds, that a decision, passed by such competent
judges, and that has now been in force for a hundred years, should all at once be reversed. Fermat has strong claims undoubtedly on the gratitude of posterity; and we do not believe that there
exists, either among the productions of ancient or
modern science, a work of the same size with his
Opera Varia, that contains so many *traits* of original invention. He had certainly approached
very near to the differential or fluxionary calculus,

as his friend Roberval had also done. He considered the infinitely small quantities introduced in his method of drawing tangents, and of resolving *maxima* and *minima*, as derived from finite differences; and, as Laplace remarks, he has extended his method to a case, when the variable quantity is irrational. He was, therefore, very near to the method of fluxions; with the principle of it, he was perfectly acquainted;—and so at the same time were both Roberval and Wallis, though men much inferior to Fermat. The truth is, that the discovery of the new calculus was so gradually approximated, that more than one had come quite near it, and were perfectly acquainted with its principles, before any of the writings of Newton or Leibnitz were known. That which must give, in such a case, the right of being considered as the true inventor, is the extension of the principle to its full range; connecting with it a new calculus, and new analytical operations; the invention of a new algorithm with corresponding symbols. These last form the public acts, by which the invention becomes known to the world at large, the judge by which the matter must be finally decided. Great, therefore, as is the merit of Fermat, which no body can be more willing than ourselves to acknowledge; and near as he was to the greatest invention of modern times, we cannot admit that his property in it is to be put on a footing with that of

Newton or of Leibnitz ;—we should fear, that, in doing so, we were violating one of the most sacred and august monuments that posterity ever raised in honour of the dead.

It has been already stated, that, when all the different ways in which an event can fall out, are equally possible and independent of one another, the fraction which expresses the probability, that the event may have certain conditions, is one which has for its numerator all the cases favourable to such conditions, and for its denominator all the cases possible. But when the event that happens affects that which is to follow, the question becomes sometimes of considerable difficulty. Laplace mentions one case, simple indeed, but important in its application. Suppose a fact to be transmitted through twenty persons ;—the first communicating it to the second, the second to the third, &c. ; and let the probability of each testimony be expressed by $\frac{9}{10}$, (that is, suppose that of ten reports made by each witness, nine only are true,) then at every time the story passes from one witness to another, the evidence is reduced to $\frac{9}{10}$ of what it was before. —Thus, after it ' as passed the whole twenty, the evidence will be found to be less than $\frac{1}{8}$.

" The diminution of evidence by this sort of transmission may," says Laplace, " be compared to the extinction of light by the interposition of several pieces of glass ; a small number of pieces will be sufficient to render an object entirely invisible which a single piece allowed to be seen very distinctly. Historians do not seem," he adds, " to have paid sufficient attention to this degradation of the probability of facts when seen across a great number of successive generations."

It does not appear, however, that the diminution of evidence here supposed is a necessary consequence of transmission from one age to another. It may hold in some instances; but in those that most commonly occur, no sensible diminution of evidence seems to be produced by the lapse of time. Take any ancient event that is well attested, such, for example, as the retreat of the ten thousand, and we are persuaded it will be generally admitted, that the certainty of that event having taken place is as great at this moment as it was on the return of the Greek army, or immediately after Xenophon had published his narrative. The calculation of chances may indeed be brought to depose in favour of it; for the probability will be found to be very small, that any considerable interpolation or change in the supposed narrative of Xenophon could have taken place without some historical document remaining to inform us of such a change. The combination of the chances necessary to produce and to conceal such an interpolation is in the highest degree improbable; and the authority of Xenophon remains, on that account, the same at this moment that it was originally. The ignorance of a transcriber, or the presumption of a commentator, may vitiate and alter a passage ; but there is a virtue in sense and consistency by which they restore themselves. The greatest danger that an ancient author runs is when a critic like Bentley is turned loose upon his text. Yet there is no fear but that, in the arguments by which he would recommend his alterations, he will leave a sufficient security against their being received.

There is an error on the subject of chance, and of cases that are equally possible, against which it is necessary to guard.

Some writers argue as if regular events were less possible than irregular, and that in the game, for example, of Cross and Pile, a combination in which Cross would happen twenty times in succession, is less easy for nature to produce than one in which Cross and Pile are mixed together without regularity. This however is not true ; for it is to suppose that the events which have already taken place, affect those that are to follow ; and this, in what relates to chance, cannot be admitted. The regular combinations happen more rarely than the irregular, only because they are less numerous. If we look for a particular cause as acting in the cases where symmetry occurs, it is not because we suppose the symmetrical arrangement to be less possible than any other ; but it is improbable that chance

has produced it, because the symmetrical arrangements are few and the asymmetrical may be without
number. We see on a table, for instance, letters
so disposed as to make the word *Constantinople ;*
and we immediately conclude that this arrangement
is not the effect of chance : not that it is less possible for chance to produce it, than any other given
arrangement of the same fourteen letters—for if it
were not a word in any language, we would never
suspect the existence of design—but because the
word being in use amongst us, it is incomparably
more probable that this arrangement of the letters
is the work of design, than of chance.

" Events may be so extraordinary that they can
hardly be established by testimony. We would
not give credit to a man who would affirm that he
saw an hundred dice thrown in the air, and that
they all fell on the same faces. If we had ourselves been spectators of such an event, we would
not believe our own eyes, till we had scrupulously examined all the circumstances, and assured ourselves
that there was no trick nor deception. After such
an examination, we would not hesitate to admit it,
notwithstanding its great improbability ; and no one
would have recourse to an inversion of the laws of
vision in order to account for it. This shows that
the probability of the *continuance* of the laws of
nature is superior, in our estimation, to every other
evidence, and to that of historical facts the best es-

tablished.   One may judge therefore of the weight
of testimony necessary to prove a suspension of those
laws, and how fallacious it is in such cases to apply
the common rules of evidence."

It sometimes happens, however, that a prevail-
ing opinion, or a prejudice, may so diminish the
natural improbability of an event, that it shall appear easily overcome by the force of testimony.

" This has happened with men of the first abilities ; and in the age of Lewis XIV., Racine and
Pascal were two remarkable examples of it. It is hu-
miliating to see with what complacency Racine, that
admirable painter of the human heart, *and the
most perfect poet who has ever been*, relates as a
miraculous event, the cure of Mademoiselle Perrier,
the niece of Pascal, and *pensionnaire* of the Abbey of Port-Royal : It is no less painful to read
the reasonings by which Pascal endeavours to prove
that this miracle had become necessary to the cause
of religion, in order to justify the doctrine of the
Nuns of that Abbey, at that time persecuted by
the Jesuits.   The young Mademoiselle Perrier,
who was then about three years and a half old, was
afflicted with a *fistula lachrymalis ;* she touched
her sore eye with a relique which professed to be
one of the thorns of the crown placed by the Jews
on the head of our Saviour, and she believed herself cured from that instant.   Some days after, the
physicians and surgeons attested the cure, and gave

it as their opinion (in which probably they were perfectly correct) that the medicines had had no effect in bringing it about.   This event, which happened in 1656, made a great noise : All Paris," says Racine, " flocked to Port-Royal.   The crowd increased from day to day ; and God seemed to take pleasure in authorizing the devotion of the people, by the number of miracles worked in that church."

The question here touched on, how far the evidence of testimony is able to overcome that which arises from our experience of the course of nature, is one of the most delicate and important which the Doctrine of Probability presents.   That testimony itself derives all its force from experience, seems very certain.   This, however, has sometimes been disputed ; and it has been urged, that there is a natural tendency to believe in the testimony of others, independent of experience.   That such a tendency really exists, we are willing to allow.   A man who feels in himself a propensity to speak the truth readily supposes a like propensity in others ; and therefore, previous to all experience, may be disposed to believe in their testimony.   He soon learns, however, that he cannot trust safely to this principle ; for he perceives, that though men have a tendency to speak the truth, they have often motives that lead them to do the contrary, that tempt them to conceal and even to pervert it ; and how much these opposite motives may counteract one another, is a matter only to be collected from experience and observation.   Indeed, it is quite evident, that whatever propensity we *naturally* have to believe in testimony, it must be in itself extremely falla- cious, as bearing no proportion to the probability of the thing believed, or the likelihood that it will hap- pen.

It is useless, therefore, in treating of probabili- ty, to talk of a tendency to believe, which, confes- sedly not being regulated by the experience of the past, cannot be depended on for its anticipation of the future.   Such a tendency, whether natural or acquired, is evidently no better than a mere preju- dice, and is as likely to lead to error as to truth. The evidence of testimony, then, is measured in the same way with other probabilities, and is ex- pressed by the number of instances in which men, circumstanced in a particular way, have been known to speak true, divided by the number of cases in which they have given evidence whether true or false.   It is true that the strict arithmetical value of this fraction is hardly possible, in any case, to be assigned.   But a certain coarse and loose estimate of it may be formed, sufficient for directing the judgment and the conduct on ordinary occasions.

The first author, we believe, who stated fairly the connection between the evidence of testimony and the evidence of experience, was Hume, in his

Essay on Miracles, a work full of deep thought and enlarged views ; and, if we do not stretch the principles so far as to interfere with the truths of religion, abounding in maxims of great use in the conduct of life, as well as in the speculations of philosophy.

Conformably to the principles contained in it, and also to those in the Essay now before us, if we would form some general rules for comparing the evidence derived from our experience of the course of nature with the evidence of testimony, we may consider physical phenomena as divided into two classes, the one comprehending all those of which the course is known from experience to be perfectly uniform ; and the other comprehending those of which the course, though no doubt regulated by general laws, is not perfectly conformable to any law with which we are acquainted; so that the most general rule that we are enabled to give, admits of many exceptions. The violation of the order of events among the phenomena of the former class, the suspension of gravity for example,—the deviation of any of the stars from their places, or their courses in the heavens, &c.—these are facts of which the improbability is so strong, that no testimony can prevail against it. It will always be more wonderful that the violation of such order should have taken place, than that any number of

witnesses should be deceived themselves, or should
be disposed to deceive others.

It is here very well worth attending to, how
much the extension of our knowledge tends to give
us confidence in the continuance of the general
laws of nature, and to increase the improbability of
their violation.   Suppose a man not at all versed in
astronomy, who considers the moon merely as a luminous circle that, with certain irregularities, goes
round the earth from east to west nearly in 24
hours, rising once and setting once in that interval.
Let this man be told, from some authority that he
is accustomed to respect, that on a certain day it
had been observed at London, that the moon did
not set at all, but was visible above the horizon for
24 hours :—there is little doubt that, after making
some difficulty about it, he would come at last to
be convinced of the truth of the assertion.   In this
he could not be accused of any extraordinary and
irrational credulity.   The experience he had of
the uniform setting and rising of the moon was but
very limited ; and, the fact alleged, might not appear to him more extraordinary, than many of the
irregularities to which that luminary was subject.
Let the same thing be told to an astronomer, in
whose mind the rising and setting of the moon
were necessarily connected with a vast number of
other appearances ; who knew, for example, that
the supposed fact could not have happened, unless

the moon had deviated exceedingly from that orbit
in which it has always moved ; or the position of
the earth's axis had suddenly changed ; or the at-
mospherical refraction had been increased to an
extent that was never known.   Any of all these
events must have affected such a vast number of
others, that, as no such thing was perceived, an incredible body of evidence is brought to ascertain
the continuance of the moon in her regular course.
The barrier that generalization and the explanation of causes thus raises against credulity and superstition,—the way in which it multiplies the evidence of experience, is highly deserving of attention, and is likely to have a great influence on the
future fortunes of the human race.

Against the uniformity, therefore, of such laws,
it is impossible for testimony to prevail.   But with
those laws that are imperfectly known, and that
admit of many exceptions, the violations are not so
improbable, but that testimony may be sufficient to
establish them.   In our own time it has happened,
that the testimony produced in support of a set of
extraordinary facts, has been confirmed by a scrupulous examination into the natural history of the
facts themselves.   When the stones which were
said to have fallen from the heavens came to be
chemically analyzed, they were found to have the
same characters, and to consist of the same ingre-
dients, nearly in the same proportions.   Now, let

us suppose two such instances :—the first person gives the stones into the hands of a naturalist, and their characters are ascertained ; the second does so likewise, and the stones have the same character. Now, if this character were one which, like that of sandstone, or of limestone, belongs to a numerous class, the chance of the agreement might be considerable, because the chance that the second observer should fall on a stone exactly of the same species with the first, would be as the number of the stones existing of that species, divided by the whole number of stones, of all different species existing on the face of the earth. This, with regard to sandstone or limestone, might be a large fraction ; and the coincidence of the two testimonies in a falsehood might not be extremely improbable. But if the species is a very rare one, the probability of the coincidence becomes extremely small. Suppose, for example, that it is a species, numerous in a medium degree ; and as there are reckoned about 261 species, let us suppose that the individuals of the species to which the meteoric stones belong amount to $\frac{1}{261}$th part of all the stones on the surface of the earth. The accidental coincidence of the second witness with the first is denoted by the fraction $\frac{1}{261}$; of a third with the other two, by

$$\frac{1}{261} \times \frac{1}{261} = \frac{1}{68121} ;$$ of a fourth with the other three,

by $\frac{1}{(261)^3}$; and so on.   As there are more than ten such cases, the chance of deceit or imposture is not more than $\frac{1}{(261)^9}$: that is, 1 divided by the 9th power of 261, or by a number so large as to consist of 22 places.   This fraction, though extremely small, is vastly greater than the truth.   The individuals of this species, instead of making a 261th part of all the stones on the surface of the earth, make, so far as we know, no part of them at all. Here, therefore, we have a testimony confirmed, and rendered quite independent of our previous knowledge of the veracity of the witnesses.

The truth of the descent of these stones on the evidence of testimony alone, would have been long before it gained entire credit ; and scepticism with respect to it would have been just and philosophical In certain states of their information, men may, on good grounds, reject the truth altogether.

The way in which probability is affected by the indefinite multiplication of events, is a remarkable part of this theory.   If out of a system of events governed by chance (or by no perceivable law) you take a small number, you will find great irregularity, and nothing that looks like order, or obedience to a general rule.   Increase the number of events, or take in a larger extent of the domain over which you suppose chance to preside, you will find the irregularities bear a much less proportion to the whole; they will in a certain degree compensate for one another; and something like order and regularity will begin to emerge. In proportion as the events are further multiplied, this convergency will become more apparent; and in summing up the total amount, the events will appear adjusted to one another, by rules, from which hardly any deviation can be perceived.

Thus, in considering the subject of life and death; if we take a small extent of country, or a few people, a single parish for instance, nothing like a general rule will be discovered. The proportion of the deaths to the numbers alive, or to the numbers born; of those living at any age to those above or below that age,—all this will appear the most different in one year, compared with the next; or in one district compared with another. But subject to your examination the parish registers of a great country, or a populous city, and the facts will appear quite different. You will find the proportion of those that die annually out of a given number of inhabitants fixed with great precision, as well as of those that are born, and that have reached to the different periods of life. In the first case, the irregularities bear a great proportion to the whole: in the second, they compensate for one another; and a rule emerges, from which the deviations on opposite sides appear almost equal.

This is true not only of natural events, but of

those that arise from the institutions of society, and the transactions of men with one another.— Hence insurance against fire, and the dangers of the sea.   Nothing is less subject to calculation, than the fate of a particular ship, or a particular house, though under given circumstances.   But let a vast number of ships, in these circumstances, or of houses, be included, and the chance of their perishing, to that of their being preserved, is matter of calculation founded on experience, and reduced to such certainty, that men daily stake their fortunes on the accuracy of the results.

This is true, even where chance might be supposed to predominate the most; and where the causes that produce particular effects, are the most independent of one another.

Laplace observes, that at Paris, in ordinary times, the number of letters returned to the Post Office, the persons to whom they were directed not being found, was nearly the same from one year to another.   We have heard the same remark stated of the Dead Letter Office, as it is called, in London.

Such is the consequence of the multiplication of the events least under the control of fixed causes : And the instances just given, are sufficient to illustrate the truth of the general proposition ; which Laplace has thus stated :—

" The recurrences of events that depend on

chance, approach to fixed ratios as the events become more numerous, in such a manner that the
probability of the mean results not differing from
those ratios by any given quantity, may come nearer
to certainty than the smallest limit that can be assigned."

Thus, if in an urn, the number of white balls to
that of black, have the ratio of p to q, the number of
white balls brought out if the whole number drawn
be n, will approach to $\dfrac{p}{p+q} \times n$, the more nearly
the greater that the number n is taken.

This proposition is deducible *a priori* from the
theory of Probability. It was first demonstrated
by Bernoulli, in the Ars Conjectandi, by a method
that is very elaborate, and confessedly the work of
much thought and study. A more simple demonstration was given by Demoivre, in his Doctrine of
Chances. Our author, in his Théorie Analytique,
has given one much preferable to either, deduced
from his Theory of Generating Functions.

The solution of another curious problem which
Laplace has given, is closely connected with the
preceding. An event having happened a certain
number of times in succession, what is the probability that it will happen once more ?

When the number of times the event has happened is small, the formula that contains the answer
to this question is considerably complicated ; when

the number is very great, it is extremely simple. Suppose the number to be n, the chance that the same event will again occur, is $\dfrac{n+1}{n+2}$, which, if n be great, is very near to unity, and may express a probability not sensibly inferior to certainty.

Thus, supposing with Laplace, that the greatest antiquity to which history goes back is 5000 years, or 1826213 days, the probability that the sun will rise again to-morrow, is, according to this rule, $\dfrac{1826214}{1826215}$; or there is 1826214 to 1, to wager in favour of that event. This, therefore, may be considered as affording a measure of the probability that the course of nature will continue the same in future that it has been in time past. It is not however on the refined principles of this calculus, that the universal belief of mankind in such continuance is founded. The above theorem was first given by Bernoulli. Our author's demonstration of it in the Essai Analytique, we believe to be new and more simple than any other.

The same multiplication of events enables us to employ the theory of probability in the discovery of causes. On this subject Laplace has made a number of very important observations. The phenomena of nature are for the most part enveloped in such a number of extraneous circumstances, and so many disturbing causes unite their influence, that it is very difficult, when they are small, to separate them from one another. The best way to discover them is to multiply observations, that the accidental effects may destroy one another, and leave a mean result containing only what is essential to the phenomenon. The entire removal of the accidental part is not to be expected, as has just appeared, without an infinite number of observations: the greater the number of observations, however, the more nearly is this mean result approximated.

Of this application of the doctrine of Probabilities, a number of examples are then given. The first relates to the diurnal variation of the barometer, as found from the observations of that instrument made at the Equator, where it is least subject to the action of irregular causes. From these, there appears to be a small diurnal oscillation, of which one *maximum* takes place about 9 in the morning, and a *minimum* about 4 in the evening: a second *maximum* at 11 at night, and a second *minimum* about 4 in the morning. The oscillations of the day are greater than those of the night, and amount to about $\frac{1}{10}$th of an inch. The inconstancy of the weather does not allow this variation to be immediately observable without the tropics, or within the range of the variable winds. Nevertheless, by applying the calculus of Probabilities to a great number of accurate barometrical observations made by Ramond during several successive years,

Laplace has found such indications of the same oscillation, as to leave no doubt of its existence,
though concealed under the irregular action of
many accidental causes. This oscillation having
its period equal to a solar day, must arise from the
sun's action, most probably, in the heating and
cooling of the atmosphere.

To the same calculus, in what regards the irregularities of the planetary system, our author pro-
fesses to be greatly indebted. The difficulty in
such cases is, often, to know whether a certain small
irregularity, combined as it is with many other ir
regularities, has an existence or not. If it has an
existence, it will give a certain determination to all
the results one way more than another; and by
comparing a great number of results, the reality of
the determination may be discovered. It is just as
if a *die* were thrown a great number of times, and
it was required to find whether it had a bias to a
certain side or not. After a vast number of throws,
if there is no bias, each face must have turned up
nearly the same number of times. If this is not
found to hold; if there be one face which has
turned up considerably oftener than the rest, it may
safely be concluded that there is a bias to that side;
and from the calculus of probabilities, the amount
of the bias may be estimated.

In this way, the calculus may be applied to several astronomical phenomena, and may be consider-

ed as a means of discovering from *induction*, some conclusions that could hardly be otherwise obtained. Laplace gives an instance of this in his own researches, concerning the diminution of a certain inequality in the precession of the equinoxes, relatively to the moon only, which was suspected by Mayer, but rejected by most astronomers as not being explained on the principle of gravitation. A scrupulous examination of observations, and the application of the calculus, convinced Laplace, that the existence of the inequality was highly probable; so that he began to look out for the cause of it. It was not long before he perceived that it must arise from the spheroidal figure of the earth, which must change a little the laws of gravity towards that body, and produce of consequence an inequality in the lunar motions. This cause had hitherto been neglected by astronomers; but, when taken into account, it explained with precision the irregularity in question, and the magnitude which, by the rules of probability, he had been led to assign to it. Other instances are given in the irregularities of Jupiter and Saturn, the satellites of Jupiter, &c. We shall only mention one result, and it is a very remarkable one, deduced from the motions of the planets being all in the same direction.

" One of the most remarkable phenomena in the solar system is, that the motions of rotation and of revolution in the planets and satellites are all in the same direction, viz. in that of the sun's rotation, and not far from the plane of his equator. A phenomenon so remarkable cannot be the effect of chance: and it obviously indicates one *general cause,* which has determined all these motions. To estimate the probability with which this cause is pointed out, it must be considered, that the planetary system, such as we now see it, is composed of eleven planets and eighteen satellites; and that the rotation of the sun, of six planets, of the satellites of Jupiter, of the ring of Saturn, and of one of his satellites, are all known. These movements, taken in conjunction with those of revolution, make a total of forty-three—all in the same direction. Now, by the calculation of probabilities, it will be found that there are more than 4 millions of millions to wager against one, that this disposition is not the effect of chance; a probability much superior to that of the historical events about which we entertain the least doubt. We must therefore believe at least with equal confidence, that One *Primitive Cause* has directed all the planetary motions; especially when we consider, that the greater part of these motions are also nearly in the same plane."

Our author proceeds, then, to offer some conjectures concerning the *physical cause* to which these motions are to be ascribed. He brings together a great number of facts, from Dr Herschell's observations concerning the *nebulæ*, which, combined with the preceding, seem to point out the solar atmosphere as the most probable cause. But where the facts lie so far out of the reach of accurate observation as many of these do, and when the supposed cause has ceased so entirely to act, the evidence we can have is so slight, and the difficulties so many, that even the author of the Mécanique Céleste must fail in giving weight and durability to his system.

In those sciences which are in a great measure conjectural, such as medicine, agriculture, and politics, the *calculus* of probabilities may be employed for discovering the value of the different methods that are had recourse to. Thus, to find out the best of the treatments in use in the cure of a particular disease, the comparison of a number of cases, where the circumstances have been as much alike as possible, will enable us to judge of the accidental causes that in each particular case assisted or impeded the cure : these last will make a compensation for one another ; and if the number of cases is sufficiently great, will leave the efficacy or inefficacy of the remedies distinctly visible.

" The same," he adds, " may be applied to political economy ; with respect to which, the operations of governments are so many experiments, made on a great scale, and calculated to throw light on the conduct to be pursued on similar occasions. So many unforeseen, concealed, and inappreciable causes, have an

influence on human institutions, that it is impossible
to judge *a priori* of their effects.—Nothing but a long
series of experiments can unfold these effects, and
point out the means of counteracting those that are
hurtful. It would conduce much to this object,
if, in every branch of the administration, an exact
register were kept of the trials made of different
measures ; and of the results, whether good or bad,
to which they have led."

He concludes with a maxim, which the circumstances of the times in which he has lived, must
have but too deeply engraven on the mind of every
Frenchman.

 " Ne changeons qu'avec une circonspection ex-
trême nos anciennes institutions et usages auxquels
nos opinions et nos habitudes se sont depuis long-
tems pliées. Nous connaissons bien par l'expé-
rience du passé les inconvéniens qu'ils nous pré-
sentent ; mais nous ignorons quelle est l'étendue
des maux que leur changement peut produire."

These are safe and just maxims ; and we are glad
to think that he who expresses them holds a high situation in the government of his country. There is,
however, another maxim grounded also on the
doctrine of Probability, which we should think
hardly less necessary than this, viz. that the rulers
of mankind, in order to remove as much as possible all chance of sudden and great revolutions,
would strike at the roots of the causes which so of-

ten render them inevitable, by taking care that all political institutions are gradually and slowly corrected, as their errors are found out, or as new circumstances in the situation of the world render them inapplicable. The negative precept, of not changing things but slowly, is not alone sufficient; it is necessary to add the affirmative precept, of changing them slowly, but readily, when reason for such change appears. In this way, the causes that tend to disturb the public order are prevented from accumulating, so as to create, or even to justify, the spirit of revolution ; and by gradual reformations, which may be made without danger, those great changes are avoided which cannot happen without incalculable mischief.

One of the most important applications of the doctrine of Probability, is to determine the most probable *mean*, or average, among a number of observations. The most accurate experiments and observations are liable to errors, which must affect the truth of the results obtained from them. To make these disappear as much as possible, observations must be greatly multiplied, in order that the errors in defect and in excess may destroy one another, and the mean, of consequence, become nearly correct. Still, however, the manner of striking this mean to the greatest advantage, remains to be examined, as also the degree of error to which, after all, it must be liable.

For a long time mathematicians were contented with taking the arithmetical mean as the true result of the observations ; that is, they added them all together, and divided the sum by the number of observations. This was sufficient when the observations appeared to be all equally good, and entitled to equal weight in the determination of the result. This, however, was far from being always the case ; and Cotes was the first, as Laplace remarks, who thought of a method by which each observation should have an influence in the determination of the results proportioned to its real value. Suppose that it is the position of an object that is required to be found by astronomical observation ; let the place given by each individual observation be found, and at each of these conceive a weight to be placed proportional to the accuracy, or inversely as the error which it is reasonable to assign to that particular observation ; the centre of gravity of all these weights is the true, or the most probable place of the object. This was in fact a generalization of the common method of taking an arithmetical mean ; for it is only conceiving, that if one observation A, was twice as good as another observation B, then, instead of A, there should be accounted two observations of the same value with B, and giving the same result with A, and so on in any other proportion, even if the proportion were expressed by a fraction. The principle here is, that after a great number of observations, the errors in opposite directions (the positive and negative errors) must be equal. This is true, if the number were infinitely great ; and, in all cases, affords a probable approximation to the truth.

The above theorem, which Cotes has given at the end of his Estimatio Errorum, admits of a simple analytical expression, but does not appear, as is remarked by Laplace, to have been made use of till Euler, in his tract on the Inequalities of Jupiter and Saturn, employed equations of condition, for the first time, in determining the elements of the orbits of these two planets. Much about the same time, Tobias Mayer employed similar methods in his Inquiry into the Libration of the Moon, and afterwards in his Lunar Tables.

The method of Cotes, when there is but one result to be determined, is of most easy application ; but when there are more than one, and, of consequence, as many equations as there are observations, it is not obvious how it can be applied, and how the equations are to be combined to the best advantage. The idea occurred to Legendre to introduce another equation, by supposing the sums of the squares of the errors of the observations to be a *minimum.* * This is a very happy generalization

* Nouvelles Méthodes pour la Détermination des Orbites des Comètes. Paris, 1806.

of the method of the centre of gravity, and applicable to cases to which it could not easily be accommodated. The same idea occurred to M. Gauss about the same time. It was not demonstrated, however, till it was done in the Théorie Analytique of Laplace, that the result thus obtained is the best of all, that which leaves the least probable error, the limits of which are assigned at the same time.

The mean result being determined, the following rule for the limit of the accuracy is given : *Take the difference between the mean result of all the observations, and the result of each particular observation. The mean error, or the greatest that is to be feared, (and it may be either positive or negative,) is a fraction, having for its numerator the square root of the sum of the squares of the differences above obtained, and for its denominator the number of observations multiplied into the square root of the number which denotes the ratio of the circumference to the diameter.*

Thus, if the differences between the mean of the observations and the observations themselves be *a*, *b*, *c*, *d*, and if *n* be their number, the mean error is

$$\frac{\sqrt{a^2 + b^2 + c^2 + \&c.}}{n\sqrt{\pi}}$$

It would be unsafe to wager that the error was less than this quantity.

It will no doubt appear singular, that a quantity $\sqrt{\pi}$ having apparently no connection with the matters in hand, should enter into the above expression. It is introduced there by the operation of integration ; by means of which, it is often brought into expressions, where it was not expected. Bernoulli was the first who found the quantity π enter into the expressions of probability ; and he appears to have thought it very remarkable.

The preceding conclusion may be useful in many cases of practical astronomy, and in other parts of natural philosophy ; or, indeed, when any thing is to be determined in quantity or position from a great number of observations ; and especially when the things to be found are represented by the coefficients of the terms of an algebraic formula.

As an instance :—Suppose it were required having two sorts of lunar tables ; and, having compared them with observations, to determine which is the best. The common way is to add together the errors of observation, and to take the arithmetical mean ; the tables to which the least mean error belongs, are accounted the best. This, however, is not the way in which the question ought to be decided. The sums of the squares of the differences between the observed and the calculated places should be added together : that set in which the square root of the sum divided by the number of observations is least, is the most exact. If the number of the terms be the same, the mere comparison of the

12

sums of the squares decides on which side the preference lies.   This instance of the utility of the
method of finding the mean is given by Laplace
himself.   Another of the same kind may be added.
—Suppose that two chronometers have been compared with the sun at noon, for a certain number
of days running, and from the register kept of
their errors it is required to find which of them is
the best.   This ought to be done by taking the
squares of the differences of the errors of the chronometer for every day ; that in which the sum of
these squares is the least, is the preferable time-
keeper.   If it is required to compute the error
that might be found, if either of them were applied to find the longitude, it will be determined
by the formula above, and will be very considerably
different from the result that would arise from a
mere arithmetical mean.

We have here an instance of a problem, to which,
in this country, very frequent recourse has been had
in the trials of chronometers for the longitude.
The only method of resolving it, has hitherto been
by finding the arithmetical mean, which, however,
the late Astronomer-Royal did in a particular way,
which, though not the same with this, was probably the best then known.   It is, however, certain,
that the true going of a clock, or the measure of
its merit, cannot be accurately determined, but by
means of the rule which has just been explained.

We shall conclude our extracts from this small but comprehensive volume, with one from the article on Population, which we have great pleasure in laying before our readers.

" The ratio of the population to the number of births would be increased if we could diminish or destroy any disease that is dangerous and common. This has been done, happily, in the case of the small-pox, first by the common inoculation for the disease itself, and afterwards in a much more complete manner by the vaccine inoculation, the inestimable discovery of Jenner, who has rendered himself, by that means, one of the greatest benefactors of the human race.

" The most simple way of calculating the advantage which the extinction of a disease would produce, consists in determining from observation the number of individuals of a given age who die of it yearly, and in subtracting the amount from the total number of deaths of persons of that same age. The ratio of the difference to the total number alive at the same age would be the probability of dying at that age if the disease did not exist. By summing up all these probabilities from the beginning of life to a given age, and taking the sum from unity, the remainder will be the probability of living to that age, on the hypothesis of the disease in question being extinguished. From the series of these probabilities, the mean duration of life on the same supposition may be computed, according to rules that are well known. M. Duvilard has found that the mean duration of human life is increased at least *three years* by the vaccine inoculation." p. 69.

But as this review is now in danger of becoming longer than the book reviewed, we shall conclude, with recommending to our readers the perusal of the work itself; and with assuring them, that they will find in it much valuable and important matter, which has not fallen within the scope of this analysis.

FINIS.

REVIEW

OF

BARON DE ZACH,

ATTRACTION DES MONTAGNES.

––––––––––

REVIEW

OF

BARON DE ZACH,

ATTRACTION DES MONTAGNES.

———

THE Baron de Zach is known in the scientific
world as an astronomer, and as the author of several works on the practical parts of the mathematical sciences. He is a native of Germany ; and his
principal residence, if we mistake not, has been at
the court of the Prince of Saxe-Gotha. He appears, from what is mentioned in these volumes,
to have been employed in 1802 by the King of
Prussia, in constructing a map of Thuringia, from
an actual survey. Several years ago he visited England ; and resided there for a considerable time.
He lived much in the family of Lord Egremont ;
and we owe to him the discovery of several unpublished MSS. of Harriot, one of the ablest and most
inventive mathematicians of the age in which he
lived. These the Baron found among the papers

---

* From the Edinburgh Review, Vol. XXVI. (1816.)—ED.

of the nobleman just named. They have since been consigned to the care of the University of Oxford; and are now, we have no doubt, in the progress toward publication.

Circumstances, of which he does not inform us, having led him to Marseilles in 1810, and induced him to make some considerable stay in that city, a climate and situation so favourable for observation naturally inclined him to undertake the solution of some of the great problems of practical astronomy. He was provided with a good apparatus ; and the research he thought of pursuing was one abundantly nice and difficult—the attraction of mountains.

It is to the discoverer of the principle of universal gravitation that we owe the first idea of such attraction, as a thing not only real, but capable of being ascertained by actual observation. Newton, in his Tract *De Mundi Systemate*, § 22, computes, that a plummet, at the foot of a hemispherical mountain three miles high, and six broad, (at the base,) would be drawn about two minutes out of the perpendicular. This suggestion was sufficient to rouse the attention of astronomers, who could not but remark, that a cause was here pointed out, which, in certain circumstances, might greatly impair the accuracy of their observations. It does not, however, appear that any one undertook to investigate the subject experimentally, till the visit made to the Andes by the French and Spanish

academicians about the year 1738. The sight of
the mountains which form so stupendous a rampart
along the shores of the Pacific Ocean, could not
but remind these astronomers of the influence which
such masses might have on the accuracy of the observations by which they were to ascertain the figure and magnitude of the earth.   M. Bouguer, a
most active and skilful astronomer, proposed to ascertain the fact by actual observation ; and began
with making a coarse estimate of the effect which
might be expected from Chimboraço, the highest
of the Cordilleras, elevated more than 3000 toises
above the level of the sea, and not less than 1700
above the level of the plain from which it rises.
From the dimensions of this enormous mass, he
computed that it might draw the plummet out of
the perpendicular by 1′ 40″ ; a quantity much too
large to escape observation.

So skilful and ingenious an observer as Bouguer,
could not fail quickly to perceive, that there were
more ways than one by which the quantity of this
attraction might be experimentally ascertained.

It is obvious that, abstracting from all disturbance of the plumb-line, the altitude of any given
celestial body when it passes the meridian is the
same in all places under the same parallel of latitude, or in all places due east and west of one another. If, therefore, two stations are chosen, one at
the foot of a mountain, suppose on the south side,

and another at a considerable distance to the east or west of the former, the meridian altitude of the same star, if the mountain have no attraction, will be the same at both these stations. But if, at the first station, the plummet be drawn towards the mountain, that is, if the apparent zenith be carried towards the south, the meridian altitudes of the star at the two stations, will differ, by the deviation of the plumb-line from the true perpendicular. If, then, observations are made at two such stations, the questions, whether the mountain has any attraction, and what the quantity of that attraction is, will both be resolved. It will add to the accuracy of the determination, if stars to the south and north of the zenith he observed at both stations. Those to the south will have their zenith distances diminished, and those to the north will have them increased, by the same quantity, when compared with the observations made beyond the influence of the mountain ; so that the effect to be measured will be doubled.

Another method proposed by these academicians was, to take two stations, one on the south, and another on the north side of the same mountain, and as nearly as possible in the same meridian. From the zenith distances of the same stars observed at each station, the difference of their latitudes might be very accurately determined. The difference of the latitudes might also be determined from the distance between the same stations, found from a trigonometrical survey of the ground. The difference of these determinations would give the sum of the deviations of the plumb-line on the opposite sides of the mountain ; and, when divided in the inverse ratio of the squares of the distances of the stations from the centre of gravity of the mass, would give the deflection of the plummet at each station. *

A third method supposes one observer to be placed at the eastern foot of the mountain, and an-

---

* The Baron de Zach has fallen into an error in quoting, or rather in interpreting, a rule laid down by Bouguer, for dividing the deflection between the two stations, and allowing to each side of the mountain its due proportion of the effect. The formula of that academician, in the case that the stations are in the same meridian, but at different distances from the centre of gravity of the mountain, requires that the sum of the deviations, or the total deviation observed, should be divided between the stations, in the inverse ratio of the squares of their distances from the centre of gravity of the mountain, as is stated above. The Baron makes it in the direct ratio of the cubes ; referring at the same time to Bouguer, whose general proposition, on the contrary, gives the result just mentioned. The theorem of the Baron is obviously wrong ; and even the theorem of Bouguer is but a coarse approximation ; as, in an irregular figure such as that of a mountain, the attraction does not vary as any power of the distance from the centre of gravity, or from any fixed point whatsoever.

other at the western. If each of these observers
regulate his clock exactly by equal altitudes, or by
the time when the sun passes over the meridian,
the difference of time pointed out by the clocks, or
the difference of longitude of the stations, will be
greater than if the mountain had not acted on the
plummets, and carried the one zenith too far to the
east, and the other too far to the west. If this difference, therefore, be determined by signals made
at each station, and observed at the other, it will
be discovered, whether the differences of longitude
so found correspond to the measured distance by
which the one observatory is east or west of the
other. This method, though perfectly good in
theory, would be found more subject to error than
that just described, in the same degree that the
difference of the longitude of two points is less
easily determined than their difference of latitude.

The French academicians made trial of the first
of these methods, by placing their instruments on
the south side of the great mountain of Chimboraço. They observed the meridian altitude of some
stars on the north, and of others on the south side
of the zenith ; and they repeated the same observations of the same stars a league and a half to the
west of the first station, where they conceived
themselves to be out of the reach of the action of
the mountain. By comparing the observations at

the two stations, they had a difference twice as great as if they had only observed stars on one side of the zenith. They would, however, have preferred the method of observing, first on the south, and then on the north side of the mountain, if it had not been that Chimboraço is inaccessible from the north. They found, in this way, that the zenith, by the action of the mountain on the plummet, had been carried $7\frac{1}{2}''$ towards the south ; a quantity vastly less than they had anticipated, and insufficient, in reality, considering that their instruments were not so perfect but that inconsistencies of $19''$, and even $26''$, sometimes entered into their observations, to determine the question whether the mountain had or had not a sensible effect on the plumb line. From that time, however, to the year 1773, no attempt was made to determine this curious and interesting fact in physical astronomy. In that year, the Astronomer-Royal at Greenwich proposed to the Royal Society of London to make an experiment of the same kind on some of the mountains of Great Britain. After a careful survey of the principal mountains both in England and Scotland, the mountain of Schehalien, * in the latter country, was judged to be more

---

* In a note on the word *Schehalien,* our author says:—
" Montagne appelée dans le pays, en langue Erse, *Maiden-pap,* qui veut dire *orage perpetuel.*" There could not be a

advantageously situated than any other. Dr Maskelyne himself undertook the operation ; and with
the same excellent zenith sector which he had carried to the island of St Helena when he went to
observe the transit of Venus in 1761, he observed
the zenith distances of stars, first on the south, and
then on the north side of the mountain. Notwithstanding a most unfavourable summer, he made
337 observations, and determined the zenith distances of the same 40 stars at each of the two stations. The difference of the latitude of the two
stations obtained from these observations, compared with that which was inferred from the measurement of their distance on the ground, gave decidedly 5″.8 for the action of the mountain on the
plummet of the sector. The great number of these

---

more unfortunate translation. The Gaelic etymologists do
indeed differ as to the derivation and import of the word
Schehalien. According to one derivation, it signifies *Maiden-
pap ;* according to another, it is said to signify perpetual
storm : And if the figure of the mountain be brought as an
evidence of the former derivation, the weather that so often
prevails around it may be brought in support of the latter.
The learned Baron, however, putting these two interpreta-
tions into one, has been so unlucky as to give *Maiden-pap*
and *perpetual storm* as synonymous expressions. From this
inaccuracy, his residence for several years in London ought
to have delivered him ; for though he could not learn there
what was Erse, he might have learned what was English.

observations, and their perfect agreement with one another, leaves no doubt at all, that mountains such as Schehalien, or of the height of 3000 feet, are able to draw the plumb-line 5″ or 6″ out of the perpendicular.

We must not here omit to observe, that the researches of the Baron de Zach have brought out a circumstance hitherto unobserved, vastly to the credit of Dr Maskelyne's accuracy. That astronomer, as he tells us himself, though he had made observations on 43 stars, did not calculate the effect from any more than the ten which he considered as the best determined, in order that he might satisfy more speedily the impatience of the Society to be made acquainted with the result of his experiments. It does not appear that he himself afterwards, or any other astronomer, ever undertook the remaining part of this task, which, however, the Baron de Zach has now performed with his usual skill and accuracy. He has calculated the results of all the 337 observations which Dr Maskelyne had made on the zenith distances of the 43 stars just mentioned. Three of these stars not having been seen from the stations both on the north and south side of the mountain, cannot be taken into account. From the 40 that remain, the conclusion deduced is, that the celestial arc between the zenith of the two observatories was 54″.651. Now, from the

measurement on the ground, the same arc comes
out 43″.019 ; the difference, 11″.632, being the
sum of the attractions of the opposite sides of the
mountain. The half of this, 5″.816, is the effect on
each side, precisely the same which Dr Maskelyne
has deduced from the observations which he considered as the best. This verification of his work
is in itself highly satisfactory, and very gratifying
to those who enjoyed the friendship, and who respect the memory of that excellent astronomer.

Among the means of resolving the problem of
the attraction of mountains, we must not omit one
which was proposed by Boscovich. This was, to
suspend a plummet from a high tower, situated on
the sea-shore, where the rise of the tide was very
great, and where the different positions of the
plumb-line, at high and low water, might be directly observed. This method, however, though
simple at the first view, is incumbered by so many
difficulties, that we believe it has never been undertaken. A very ingenious improvement on it,
proposed by the late Professor Robison, consisted
in observing the alteration of the level of a fluid,
caused by the access and recess of the great wave
of the tide, which alteration was to be measured by
the reflection of a fixed object from the surface of
the fluid. The fluid might be the water in a deep
well, close to the sea-shore. We do not think

that this notion is entirely inapplicable to practice ; and we believe all must agree that it is very ingenious.

The Baron de Zach, when at Marseilles in the year 1810, finding himself, as has been said, in a situation most favourable to astronomical observations, and being also furnished with good instruments, though not such as the zenith sector employed by Dr Maskelyne, yet conceived that the position of Marseilles, with a chain of hills rising on the one side, and the Mediterranean stretching out on the other, afforded great conveniency for trying whether, with such instruments as he possessed, the attraction of a mountain of moderate size could be rendered sensible.

The scene of his observations was the bottom of a chain of calcareous mountains, which, at the distance of 6000 or 8000 toises from the city of Marseilles, extends from east to west. The highest part of the chain, called the hill of Mimet, has an elevation of about 400 toises above the level of the sea. On the side of it, and at the height of about 250 toises, are the ruins of an old convent, known by the name of Notre Dame des Anges, commanding a fine view of the Mediterranean at the distance of five or six miles, extending indefinitely toward the south. In the south-west, at the distance of about 8000 toises from the coast, an insulated rock, in the middle of the sea, rises just above the sur-

face, and is called l'Isle de Planier, on which stands
a light-house. The position of these two points
seemed very favourable for the proposed experiment, which was to be made by determining the
difference of latitude of the two points, the convent
and the light-house, by astronomical observation,
and then connecting them together by a series of
triangles, in order to ascertain the same difference
by trigonometrical measurement. At N. D. des
Anges the hill of Mimet would exert its full action on the plumb-line, or on the liquor in the spirit level. At Isle de Planier, on the contrary, on
the surface of an insulated rock, and at the distance
of 8000 toises from the land, and 16,000 from the
mountain, the action of the hill must amount to
nothing ; and, consequently, the difference between
the amplitude of the arch of the meridian, determined by celestial observation, and inferred from terrestrial measurement, would give the effect of the attraction, uncombined with any other force either assisting or opposing it. The instruments with which
the Baron was furnished, were, a repeating circle of
twelve inches radius by Reichenbach, with which he
proposed to measure the distances from the zenith; a
repeating theodolite of eight inches radius by the
same artist, for observing azimuths and terrestrial
angles ; an English sectant of nine inches radius
by Troughton, for taking corresponding altitudes,
to regulate four chronometers, three constructed

by Josiah Emery of London, and one by Louis
Berthoud of Paris. With these he began his observations at the station of Notre Dame des Anges ;
and, by 874 altitudes, determined the true zenith
distances of a great number of stars, all which was
done between the 11th and 24th of July 1810.

The situation of this station did not allow the
observation of stars on both sides of the zenith, as
the mountain rose very perpendicularly to the north
of the convent. Such stars might indeed have been
observed with an instrument, like the sector, calculated for making observations near the zenith. The
repeating circle has not that advantage ; for its
perpendicularity to the horizon not being very accurately ascertained, and the error arising from
that source being greatest near the zenith, the instrument is ill adapted to the observations which,
in such a case as the present, would have been the
most eligible. The altitudes, therefore, observed,
were of stars considerably distant from the zenith,
and all of them to the south. Indeed, though Baron de Zach appears to be well pleased with the
advantages which both the locality already described, and the instruments that have been enumerated carried with them, we do not think that in either they were remarkable ; and if the result, after
all, has turned out favourable, it is more to be attributed to the skill, diligence, and accuracy of the
observer, than to the particular advantages which

he enjoyed. In one thing, indeed, we cannot but admire the power which an astronomer derives from the fine climate of Marseilles, compared with that of our island. Dr Maskelyne, in a residence on the side of Schehalien of four months, could hardly find the means of placing his sector in the meridian; and, with all that patience or industry could perform, could only make 337 observations. The Baron de Zach, in 13 days, on the shores of the Mediterranean, was able to make 874 observations. That Greenwich should afford, nevertheless, a greater number of observations to be completely depended on, than any other observatory in Europe, is a strong instance how, on some occasions, the moral causes can control the physical.

The data necessary in this way to determine the attraction of Mimet, required observations to be made for finding the difference of latitude between the two extreme points already mentioned; and this must be done, as has been said, not only by celestial observation, but by terrestrial measurement. For these purposes, a great number of observations were made, which the Baron has given, not only in their original state, but also as reduced and prepared for the final calculation, with a degree of order and correctness altogether exemplary. We have never seen any work of the same kind, where there is more method and order in the arrangement, more accuracy in the detail, and more fairness in striking the mean, where there is any difference among the observations. It is a book, for these reasons, which no one, engaged in similar pursuits can study with too much care.

In his discussion concerning the merit of the instruments employed in his observations, a fact occurs concerning the repeating circle, which is certainly of importance ; and, to us who are but little acquainted with the nature of that instrument, seems difficult to be explained. It appears, that these circles are subject to certain variations or anomalies, which may extend to three or four seconds, from causes altogether unknown. Our author tells us, that he had formerly remarked, in a letter addressed to the Editors of the Bibliothéque Britannique, that one cannot answer, within three or four seconds, for latitudes inferred from a long series of observations agreeing well with one another, and made with the same repeating circle ; for another circle will offer another series of observations, agreeing as well with one another, but differing constantly from the first series by 3″ or 4″. This remark, when first made by Baron de Zach, appears to have drawn upon him a good deal of animadversion, though the fact itself was not disputed.

" They have made it," says the Baron, " a kind of reproach, that I had not pointed out the precise source of these variations. My answer was not

ready, but will appear in due time.   In the mean-
while, I have the satisfaction to think, that I have
awakened and directed the attention of astronomers
and of artists, to an important point which requires
their attention."

He goes on to remark, that this defect, from
whatever cause it may arise, had no chance of affecting the determination of the attraction of the
mountain which was the present object of research.
It was not the absolute latitude, either of N. D.
des Anges, or of l'Isle de Planier, that was now required ; but it was the difference between them,
which the constant error above described could
have no tendency to affect.

The preceding remarks, however, cast a little
uncertainty on the determinations made by these
most commodious and useful instruments.   We
have certainly no right to offer any opinion about
an anomaly, which those who are best acquainted
with the subject seem hitherto at a loss to explain.
It has always appeared to us, that the smallness of
the telescopes with which instruments held in the
hand must be provided, is a considerable defect,
and may perhaps have given rise to the inconsistency just mentioned.

The second article in this Treatise, relates to the
difference of longitude between the two stations,
and contains several remarks of great value to those
engaged in similar pursuits.   Having regulated

his time-keepers at N. D. des Anges, by a series of observations of equal altitudes of the sun, the Baron proposed to determine the difference of time at that point and l'Isle de Planier, by signals made at the one station, and observed at the other. He enumerates, however, before describing these, the various ways in which the problem of ascertaining the difference of longitude of places had been attempted to be resolved. One is, as is well known, the occultation of stars by the moon, which, of all the methods purely astronomical, is certainly to be considered as the best. Yet if the star is small, and if it disappears behind the enlightened part of the lunar disk, it is often lost sight of from the comparative weakness of its light, before it actually touch the limb of the moon ; and so also, at the emersion, it has perhaps got to some distance from the moon, before it can be distinguished. It is also remarked, that, in occultations, it sometimes happens that the star, after having touched the luminous disk of the moon, still appears for some seconds upon the disk. At first it seems to advance, and afterwards disappears altogether. This is known to have been experienced by several of the most eminent observers ; by Cassini, Delahire, Feuillée, &c.

Another cause, he adds, which may render such observations defective, is the moon's parallax; in consequence of which, the immersions and emersions of stars are made at different points of the limb for observers placed in different situations. Now, it is certain that the surface of the moon is unequal, and that there are mountains on it which, according to the observations of Messrs Herschell and Schröter, are not less than 4000 toises in height. With good telescopes, one may see the little asperities which their summits form on the limb of the moon.

" It was thus that, in observing an occultation of λ Piscium on the 8th of September 1786, the star appeared to sink in the interval between two of these summits on the moon's limb, and disappeared in the hollow or valley between them. An observer, in another place of the earth, might have seen it hid behind one of these summits; and the two occultations supposed to have been made by the same part of the moon might have differed by several seconds. A singular observation, made by M. Koch, an astronomer of Dantzig, on the 7th of March 1794, on occasion of an occultation of Aldebaran, shows the effect which the mountains in the moon may produce in such cases. The star which just grazed as it were along the limb of the moon, was three times eclipsed by the mountains, before it totally disappeared behind the real disk. The immersion was near the superior horn; the star first disappeared, and in 10″ appeared again in all its brightness; after some seconds, it disappeared, and re-appeared a second time. It was soon after concealed for the third time by a mountain, but appeared once more before the real immersion behind the true disk of the moon."

However excellent, therefore, the method of occultations may be for great distances, it is insufficient for affording the necessary accuracy when the distances are small. An error of 1″ or 2″, which, on an arch of the meridian of several degrees, might be counted as nothing, would become very considerable for a difference of longitude which was only a few seconds. In such cases, the celestial signals must be abandoned, and we must have recourse to such as we can make ourselves on the surface of the earth. The first person who attempted this was Picard, in determining the difference of meridians between the observatory at Copenhagen, and the ruined observatory of Tycho Brahe in the island of Huena. He kindled a fire on the tower at Copenhagen; but he does not tell by what means he made it disappear suddenly. Other methods have been since followed. A trial of the method of finding the difference of time, by means of signals, was made near London in 1775, with great success. The signals were made by the explosion of rockets in the air, which were thrown up from Loampit-hill, near London, where Mr Aubert, a well-known lover of astronomy, had his observatory. Dr Maskelyne observed the explosion from the

Royal Observatory; Mr Wollaston from Chisle-
hurst in Kent; Mr Heberden from Pall-Mall in
London; and Mr Ellicot from Horseley Lane.
The longitudes of these five places were thus determined with the greatest precision; the differences at any one place not exceeding a fraction of
a second. The greatest distance, however, between
these places, was not more than 6 English miles,
or 3 French leagues. In the case of greater distances, the same method probably could not be
practised with equal success.

"It seems singular," the Baron observes, " that
for making such signals they have not long since
had recourse to the most natural and simple expedient, and the most easy of execution withal, that
of kindling a small quantity of gunpowder in the
open air. This signal is the most visible and the
most instantaneous that can be conceived. It is
seen at all seasons, and across rain and fog, even by
the naked eye. The sudden flash of the gunpowder
strikes the eye, though it be not directed precisely
to the point from which the light comes, and even
when the place from which the signal is made is
under the horizon of the observer. It is not only
during the night that these signals may be made,
but they may be seen in broad day, with telescopes
directed to the place where the signal is made, as I
have often experienced; and have, by that means,
been relieved from the necessity of passing the

night in bivouac, in the open air, as I must otherwise have done. The first use that was made of
this method was in the year 1740, by Cassini and
Lacaille, in measuring two degrees of longitude
near Jet, in Languedoc, and Aix in Provence.
These two stations are distant about 40 leagues.
Towards the middle of that distance they took a
station on the sea side, near the mouth of one of
the branches of the Rhone. There, from a ter-
race on the roof of a church, they set fire, evening
and morning, to 10 lib. of powder. The flashes
were seen distinctly at both extremities of the line;
and the difference of longitude concluded accord-
ly. Cassini proposed to do the same for determining the difference of meridians between Paris and
Vienna; but his proposal has never been executed.
The quantity of gunpowder which these academicians used, was much too great; and, beside the
useless expence, the signals so made were more uncertain and less instantaneous. Even with a single
pound of powder, I have observed that the flame
lasted for $2''$ or $3''$; and on that account I have
never used above 6 or 8 ounces. In 1803, I made
these signals on the Brocken, one of the highest
mountains of the Hartz, 535 toises above the level
of the sea; and the signals were seen at the distance
of more than 50 French leagues all round. What
is most extraordinary is, that they were seen at the
distance of nearly 55 French leagues on the small

hill of Keylenberg, not more than 200 toises in height; and from which the Brocken itself could not be seen, on account of the curvature of the earth. The light therefore seen at Keylenberg was nothing but the repercussion of the light of the signal from the clouds, of which it is known that there are many instances."

These remarks may be very useful to those who are engaged in similar operations, and particularly in the measurement of the degrees of longitude, or for the measurement of arches perpendicular to the meridian.

All this is followed by a table of the observations made by help of these signals, for determining the difference between the time at Nôtre Dame des Anges, and the Imperial Observatory at Marseilles. The observations were made between the 11th and 21st of July, and amount to 64. The greatest difference among them hardly exceeds $1''$; and the mean of the whole is $29''.95$ of time, or $7'29''1'''$ in degrees.

The use of the observation of azimuths for the same purposes, is considered at great length. The system of triangles was *oriented*, that is, its position in respect of the meridian ascertained by azimuths, determined chiefly from the sun's passage over the meridian, or, such as are here called, circum-meridian azimuths. The Baron afterwards recommends the method of ascertaining the azimuths by

the polar star, after the manner first employed by
General Roy, and since followed by those who
have succeeded him in the conduct of the trigonometrical survey of England. He says, that the excellent repeating theodolites constructed by Reichenbach, are well adapted to these observations;
and he gives two examples from azimuths observed
at Munich, where the angle was repeated a prodigious number of times with very small variations.
It would seem, therefore, that this theodolite carries
a telescope with a very accurate vertical motion,
though less accurate than that of Ramsden's great
theodolite. If this advantage is conjoined with the
power of repetition, it must no doubt render the
theodolite the most perfect instrument that has yet
been employed in such operations as we are now
treating of.

The measurement of the base, which was to serve
as the foundation of the trigonometrical survey,
comes next; and occupies a considerable part of the
first volume. In all the parts of this very essential
work, the greatest care seems to have been taken,
and no precaution omitted, that the skill and experience of this very ingenious astronomer could
add to the methods invented and executed by those
who had gone before him.

In the end of the volume, it appears that the difference of meridians between N. D. des Anges
and Planier, determined astronomically, is 15′

$35''.79$; and that the same, determined geodetical-ly, is $15'\, 46''$. There is a difference here of $10''.67$, which, however, does not at all affect the difference of latitude.

The result of the whole, after every possible check was introduced, is, that the astronomical ob-servations at N. D. des Anges made the latitude of that station less by $2''$ and a small fraction, than when the same was ascertained by the intervention of terrestrial measurement from the latitude ob-served at Planier. The same difference of $2''$ was deduced from the latitudes of three other stations, all so distant as to be out of the reach of the attrac-tion of the ridge of Mimet. No doubt could therefore remain, that these two seconds arose from the zenith of N. D. des Anges being carried that far south by the attraction of the mountain. It is thus very satisfactory to know, that, even with small instruments, so important a point can be settled as the determination of the attraction of a mountain—the most beautiful and most palpable verification of the law of gravitation which science has yet afford-ed. The further researches, that lead to a com-parison between the density of such a mountain and the density of the earth itself, require a num-ber of additional data, which cannot be ascertained with tolerable accuracy, but in the case of moun-tains of considerable magnitude, and as much as possible insulated. This, accordingly, the Baron

de *Zach* did not attempt ; and the only investigation of the kind yet existing, is that which was
founded on the experiments made at Schehalien.

Great as the skill and accuracy were with which
those experiments were conducted, the attraction
of mountains is a subject by no means exhausted ;
and it were greatly for the interest of science that
these experiments should be repeated under as
great a variety of circumstances as can easily be
attained. The northern part of the island of Great
Britain is well accommodated to such observations,
and the continuation of the trigonometrical survey
which is now extended to that country, affords the
best opportunity for carrying such experiments into execution. They would indeed make but a
small deviation from the general plan of the survey. Were the method to be followed that was
pursued in the survey which is the subject of those
remarks, any mountain, or chain of mountains,
having a plain of considerable extent, either to the
south or to the north, might very well be used for
determining the attraction. The observations made
in that way, though they do not double the effect,
as was done in the case of Schehalien, are so much
easier to be made, and may of course be executed
in so many more instances, that on the whole they
may be reckoned preferable. The survey of the
mountainous tract, and the gauging, as we may
call it, of the mountains, would require to be con-

tinued so far as to reach the limits beyond which
no inequality of ground can be supposed to act sensibly on the plumb-line. The Grampian mountains would afford many situations well accommodated to observations of this sort. The opposite
sides of a valley also, as, for instance, of Loch Tay,
or Loch Ness, might be used in the same way.
The two zeniths would there be made to approach
one another ; and the arch between them, found
by celestial observation, would be less than the
same concluded by trigonometrical measurement,
by the sum of the attractions of the mountains on
the south and north, *minus* that of the intervening
water, which, as a lighter substance, would have less
action on the plummet than an equal volume of
earth or rock. What related to the nature of the
rocks, would be readily ascertained by the skilful
mineralogist who is now so properly connected
with the execution of the trigonometrical survey.
It is a great additional argument in favour of what
is here proposed, that a long series of similar operations has prepared observers admirably calculated
for the present purpose. Men accustomed to live
in the open air, and encamped on the sides or the
summits of mountains, to watch the motions of the
stars for months together, and to endure all the
suffering and disappointment which the vicissitudes
of the weather inflict on none so severely as on the
astronomer. Men trained in this manner are not

often to be met with : so much experience and skill
in the nicest observations of science, can but seldom be combined with the hardiness of rural, we
might almost say, of savage life.   It were therefore
to let slip a most favourable occasion for promoting
the interests of science, not to take this opportunity of inquiring farther into the attraction of mountains.   The instruments are already on the spot,
as well as the hardy, experienced, and skilful observer who is to use them ; so that the same thing
can never be undertaken at so little expence to the
public, and in a manner so truly economical, and
so highly advantageous to science.

As an additional reason for including the inquiry
into the attraction of mountains, in the plan of the
trigonometrical survey, we must be permitted farther to state, that there are several circumstances
in the experiments at Schehalien, which should
render the repetition of them extremely desirable.

Though nothing could easily be added to the
accuracy of the astronomical part, of which we
have just now seen the strongest and most impartial evidence, yet equal praise cannot be bestowed
on the trigonometrical survey, by which the magnitude and figure of the mountain were determined.   The theodolite employed was but an imperfect instrument ; it gave the angles to minutes

only : it was furnished with telescopes of a very
moderate magnifying power ; and, though the
work of Ramsden, was in all respects inferior to
the instruments now employed for the like purposes.
Mr Burrowes, into whose hands this part of the
work was committed, was new in the employment ;
and, though skilled in mathematics and astronomy,
had no experience in the sort of work he was employed now to conduct.   As to all, therefore, that
relates to the density of the earth, and the conclusions grounded on the figure and magnitude of the
mountain, it must not be supposed that the same
precision is to be found as in the determinations
purely astronomical.

We are enabled to state this with the more confidence, that circumstances have led us to study
the detail of this survey with a more minute attention, than has probably ever been done by any one
except Dr Hutton, who has so ably conducted the
computations grounded on it.   In this examination we have remarked, that when the solid content of the mountain is reduced into columns of
equal attraction, according to Dr Hutton's method,
owing to some imperfection in the survey, the
lengths of those columns cannot always be accurately ascertained ; and, particularly when they come
nearly to the level of the observations, that it is
often uncertain whether they rise above that level,

or fall short of it, and, of consequence, whether a certain quantity is to be applied as an augmentation or a diminution of the whole attraction.

There were even faults in the plan, no less than in the execution of the experiments. The observatories were placed too high on the sides of the mountain ; they were about half way up ; so that between a sixth and a seventh of the total effect of the attraction was lost. The sections were vertical, and carried at random, some entirely, but many of them only partially across the mountain, instead of being conducted horizontally round it, and connected together by two vertical sections at right angles to one another.

In the distance to which the survey extended, no principle seems to have been adopted as a guide, except a very insecure one, that at the distance of a mile and a half, or two miles, the action of a mountain of ordinary size could not sensibly affect the direction of gravity. The knowledge obtained from the experiments at Schehalien afford a much better, and more secure principle for fixing the limits within which the attraction of a great mass of matter may be supposed to produce a sensible effect.

Add to this, that at the time of these experiments no attention, or next to none, was bestowed on the structure of the mountain, and the distribution of the materials which compose it. This omission, accordingly, gave no inconsiderable degree of vagueness to the conclusions deduced concerning the density of the earth.

It is true, that two gentlemen, zealous to contribute to the accuracy of this interesting inquiry, endeavoured, not long ago, by a mineral survey of Schehalien, to remedy this defect, and to ascertain, with some degree of precision, the specific gravity of the rocks which compose that mountain. They succeeded, perhaps, as far as the nature of the thing will now admit ; but certainly much less, than if a mountain of simpler structure had been the subject of examination, or if the mineral survey had been undertaken along with the trigonometrical, when the instruments of observation were on the spot, and all the stations distinctly recognised.

These circumstances, though they go no farther than to render the limits within which the accuracy of the results are contained, more distant than they would otherwise have been, are certainly to be held as good grounds for wishing to have the same experiments repeated, with an attention to all the improvements that have been made since the time when they were instituted. The opportunity, then, that now presents itself, we hope, will not be overlooked, when the instruments, as has been said, are prepared, and when observers are at hand, zealous to engage in the work, instructed in all the resources of their art, and accustomed to overcome all the difficulties of their situation. Such an enterprise would form a very noble conclusion of the present survey; and would distinguish it from all others yet made, as much for the variety and importance of the objects contained in the plan of it, as for the perfection of the execution. It is already infinitely to the credit of the country, and those entrusted with the government of it, that, during the long and expensive war in which the nation has been involved, this great work of science has been carried on as in the midst of profound peace. We may therefore hope, that the termination of an arduous contest, and the restoration of tranquillity to the world, will permit this national work to be completed with an extent and accuracy worthy of the spirit with which it has been begun and carried on.

FINIS.

REVIEW

OF

KATER ON THE PENDULUM.

REVIEW

OF

KATER ON THE PENDULUM *

———

THE end of the last century, and the beginning of
the present, have been distinguished by a series of
Geographical and Astronomical measurements,
more accurate and extensive than any yet recorded
in the history of science. A proposal made by
Cassini in 1783, for connecting the Observatories
of Paris and Greenwich by a series of triangles, and
for ascertaining the relative position of these two
great centres of Astronomical knowledge by actual
measurement, gave a beginning to the new operations. The junction of the two Observatories was
executed with great skill and accuracy by the geometers of England and France : the new resources
displayed, and the improvements introduced, will
cause this survey to be remembered as an Era in
the practical application of Mathematical science.

* From the Edinburgh Review, Vol. XXX. (1818.)—ED.

A great revolution had just begun to take place in the construction of instruments intended for the measurement of angles, whether in the heavens or on the surface of the earth; and was much accelerated by the experience acquired in this survey. One part of this improvement consisted in the substitution of the entire circle for the quadrant, the semicircle, or other portions of the same curve, as the unity and simplicity of the entire circle, distinguish it above all figures, and give it no less advantage in Mechanics than in Geometry. Circular instruments admit of being better supported, more accurately balanced, and are less endangered from unequal strain or pressure, than any other. The dilatation and contraction from heat and cold, act uniformly over the whole, and do not change the ratio of the divisions on the circumference.

A geometrical property of the same curve contributes also much to the perfection of those instruments, in which the whole circumference is employed; and though it be quite elementary, and has been long known to geometers, it was first turned to account by artists about the time of which we now speak. The proposition is, that two lines intersecting one another in any point within a circle, cut off opposite arches of the circumference, the sum of which is the same as if they intersected one another in the centre. Hence it follows, that, in a circular instrument, whether the centre about which the index turns be the true centre or not, the mean of the two opposite arcs is the exact measure of the angle to be found. This gives a complete correction for one of the great sources of inaccuracy in the construction of mathematical instruments, since, by opposite readings off, the error in the centering is always corrected. Ramsden, to whom the art of constructing mathematical instruments owes so much, was the first among modern artists who made an astronomical circle of considerable size. A theodolite, also, which he made for General Roy, who conducted the survey just referred to, was, of its kind, the most perfect instrument yet constructed, and was furnished with the best telescope that had been employed in geodetical observations.

In France, also, the entire circle was introduced, and with a great additional improvement, that of repeating or multiplying the angle to be measured any required number of times. The consequence of this is, that the *mean* taken by dividing the multiple angle at last obtained by the number of the repetitions, gives the angle with an exactness which would have required a great number of observations, and a great length of time, if other instruments had been used.

The first idea of this excellent contrivance occurred to Tobias Mayer of Gottingen, whose name is so well known in the history of Astronomy. The

instrument was afterwards reconstructed and highly improved by the Chevalier Borda.   In 1787,
when the Astronomers of Paris met those of England toward the conclusion of the survey, they were
furnished with repeating circles, which was the first
time that this instrument had been employed in similar observations.

As an evidence of the increased accuracy now
obtained, it may be observed that it was in the
survey of the ground between Greenwich and
Dover that the excess of the angles of a triangle a-
bove two right angles arising from the curvature of
the surface on which the angles were observed,
first became an object of actual measurement.   On
this quantity which has been called the *spherical
excess*, and was measured also by the repeating
circle, Legendre, with the ready invention that
easily accommodates itself to new circumstances,
grounded an admirable rule for reducing the solution of small spherical triangles under the power
of plane trigonometry.   The accuracy now expected was such, that an error of as many seconds in
the measure of an angle as was formerly allowed of
minutes, was no longer to be tolerated.

To Great Britain, the operations now entered
on were attended with a further advantage, Government having been induced to continue a work
so auspiciously begun, by extending a trigonometrical survey over the whole island, so as to ascer-

tain its topography with more precision than had yet been done with respect to any tract of equal extent on the surface of the Earth. The survey has accordingly been continued to the present time, and is now carrying on in Scotland under the able direction of Colonel Mudge, and by the meritorious exertions of Captain Colby, an indefatigable and accurate observer, instructed by much experience, and supported by a zeal and firmness of which there are but few examples.

It was not long after the commencement of this survey, that a system of trigonometrical and astronomical operations of still greater extent was undertaken by the French government.

The want of system in the weights and measures of every country ; the perplexity which that occasions ; the ambiguous language it forces us to speak ; the useless labour to which it subjects us, and the endless frauds which it conceals, have been long the disgrace of civilized nations. Add to this, the perishable character thus impressed on all our knowledge concerning the magnitude and weight of bodies, and the impossibility, by a description in words, of giving to posterity any precise information on these subjects, without reference to some natural object that continues always of the same dimensions. The provision which the art of printing has so happily made for conveying the knowledge of one age entire and perfect to

another, suffers in the case of magnitude a great and
very pernicious exception, for which there is no remedy but such reference as has just been mentioned. Philosophers had often complained of these
evils, and had pointed out the cure : but there
were old habits and inveterate prejudices to be
overcome ; and the phantom of innovation, even
in its most innocent shape, was sufficient to alarm
governments conscious that so many of their institutions had nothing but their antiquity to recommend them. At the commencement of the French
Revolution the National Assembly was avowedly
superior to the last of these terrors, and the philosophers of France considered it as a favourable opportunity for fixing, with the support of Government, a new system of measures and weights, on
the best and most permanent foundation.

Of the quantities which nature preserves always
of the same magnitude, there are but few accessible
to man, and capable at the same time of being accurately measured. The choice is limited to a
portion of the earth's circumference, or to the
length of the pendulum that vibrates a given number of times in the course of a solar or sidereal
day, or any portion of time accurately defined by
some of the permanent phenomena of nature. The
choice of the French mathematicians fell on the
first of these, and was accompanied with this great
benefit to science, that it enforced a very diligent

11

and scrupulous examination into the magnitude and figure of the earth. The quadrant of the terrestrial meridian was the unit of linear extension which they proposed to assume, and the ten-millionth part of it was the standard to which all linear measures were to be referred. The series of difficult and nice observations undertaken with a view to this improvement, carried on in the midst of much intestine disorder with signal firmness and perseverance, and finished, in spite of every obstacle, with all the accuracy that the new instruments and new methods could afford, has raised to the men of science * engaged in it, a monument that can never be effaced. The meridian of Paris continued to Dunkirk, on the one hand, and Solieure on the other, and afterwards extended beyond the latter to the southernmost of the Balearic Isles, amounting nearly to an arc of 12 degrees, afforded means more than sufficient for computing the quadrant of the meridian, and thus fixing the standard on sure and invariable principles.

In consequence of this, the figure, as well as the magnitude of the earth, came to be better known than they had ever been before, because of the new *data* afforded for entering into combination with the lengths of degrees already measured in different countries. The extent of the arc of the meri-

---

* Delambre, Mechain, Biot, Arago.

dian, thus determined, is also about to receive a great increase, by the addition from the British survey, of an arc extending from the parallel of Dunkirk to that of the most northerly of the Shetland Isles ; so that the distance between this last parallel and that of Formentera, nearly a fourth part of the quadrant of the meridian will become known by actual measurement.

But while it is possible to interrogate nature in two different ways concerning the same thing, curiosity is not to be satisfied without having both her responses. The pendulum, as is well known, affords the means of determining, not indeed the magnitude, but the figure of the earth ; that is, its compression at the poles, or the oblateness of the spheroidal figure into which it is formed. At the equator, gravitation is weaker than at the poles ; both on account of the centrifugal force which is greatest at the former, and vanishes altogether at the latter, and of the greater distance of the circumference of the equator from the centre of the mass. If the earth were quite homogeneous, Newton demonstrated, that the same fraction, viz. $\frac{1}{250}$ would denote the oblateness of the earth, and the diminution of gravity from the pole to the equator. There is, however, good reason to believe, that the earth is very far from being homogeneous, and is much denser in its interior than at its surface. Clairaut, therefore, did an unspeakable service to this branch of science, when he showed, that in every case the two fractions just mentioned, though not equal to one another, must always, when added together, constitute the same sum, that is, $\frac{2}{230}$, or $\frac{1}{115}$. Hence the oblateness appearing from the measurement of degrees to be $\frac{1}{312}$, the increase of gravity from the equator to the poles, or, which is the same, the shortening of the pendulum, must be $\frac{1}{182}$. We must have recourse to experiment, then, to discover, whether this be agreeable to the fact, or whether evidences thus brought together from such different regions, conspire to support the same conclusion. Laplace, accordingly, from an examination of 37 of the best observations made in different latitudes, from the equator as far as the parallel of 67 degrees, had obtained a result that agreed very well with the conclusions from the measurement of degrees. But these observations had been most of them made long ago, before the present extreme precision was introduced, and even before the means of comparing the lengths of two *rules*, or two rods of wood or of metal, was completely understood. It was therefore extremely desirable, that a series of new observations of the same kind should be made in different countries. The National Institute had begun the series at Paris : it had made a part of the Système Métrique, to determine the relation between the seconds pendulum and the *mètre* ; and a number of experiments for that purpose were made by Borda and Cassini, with every precaution that could ensure exactness.

After quiet was restored to Europe, England had leisure to attend to other objects than those in which the ideas of defence or of conquest were concerned.   France and a great part of the Continent had adopted the scheme of uniform measures; in England a plan for the same had been often thought of; it had been more than once undertaken, but never on a right system; and had always fortunately, though perhaps weakly, been abandoned. It was now begun apparently under better auspices; a bill for the purpose was brought into Parliament; and our readers may remember, that it was thrown out in the House of Peers by the opposition of a noble lord, more remarkable for the ingenuity than the soundness of his opinions.   It happened here, however, as appears to us, that his lordship was entirely in the right; the bill was a crude and imperfect scheme, prepared without due consideration of the various bearings of so nice a question, and consulting partial or present conveniency at the expence of permanent and general utility; having withal no dependence on any of those magnitudes which nature herself has taken pains to secure against vicissitude and change.

The attention of the men of science about London was now naturally turned to the experiments by which the length of the pendulum may be accurately determined. The nature of the apparatus best fitted for that object is by no means obvious. The French academicians, just referred to, had indeed employed a very simple one, which seems capable of great exactness. It consisted of a ball of platina suspended by a fine wire, and vibrating about a knife edge, which served as its axis. The vibrations counted by the person who conducted the experiment, were compared with those of a clock, placed close by, and regulated according to mean solar time. After a sufficient number of such comparisons, the length of the pendulum from the knife edge to the centre of oscillation of the ball, was partly measured and partly calculated; and thus the quantity required was determined.

Though this method is susceptible of great accuracy, and, in the hands of such men as Borda and Cassini, could not fail to lead to a satisfactory conclusion, yet it is right to have so important an element in our researches as the length of the pendulum, or the intensity of gravitation, ascertained by experiments made with different instruments; made according to different methods, and particularly not so dependent on the mathematical theory of the centre of oscillation as to be without the possibility of verification by experiment. It must not be supposed, that in laying down this last condition, we mean any thing so absurd as to question the force of mathematical demonstration. A conclusion purely mathematical, when applied to an object that is also purely mathematical, one that partakes of the same immaterial and impassible nature with itself, is above receiving additional evidence from any source whatever, and despises alike all attempts to increase or diminish its authority. But the same is not exactly the case when the conclusion is applied to a material body ; it then partakes of the imperfection of the subject ; and thus, in a sphere even of gold or platina, the actual centre of oscillation may not coincide to the ten thousandth part of an inch, with the point which the calculus has determined. In such instances the verification by experiment, if it cannot be called necessary, is at least highly satisfactory.

Among the mathematicians who endeavoured to resolve the problem on a principle of this kind, the author of the paper which is the subject of this article, came soon to be particularly distinguished. Captain Kater, to the profession of a soldier, seems early to have united the pursuits of science, and to have acquired uncommon skill and accuracy both in philosophical experiment and astronomical observation. We understand, that, in India, when a very young man, he assisted Colonel Lambton in the trigonometrical survey of Hindostan, and was extremely useful in a very nice and important part of the work, the selection of the stations where the observations were to be made, and of the summits to be intersected, a matter which requires great judgment; one which, in a mountainous country, and under a vertical sun, must be full of difficulty and danger, and from which we have been sorry to understand that his health had materially suffered.

Captain Kater having returned to England, and resumed the pursuits of science, began to consider how the experiment of the pendulum might best be made in a way to admit of verification by a reverse experiment; and a cylindric rod of brass or of iron readily occurred to him as a body well adapted to that purpose. The impossibility, however, of finding a rod or bar of metal so homogeneous that its centre of oscillation could be determined merely from its dimensions, made him quickly despair of succeeding by such means. It happily occurred to him, in this uncertainty, that there was one property of the centre of oscillation by which its place might be made manifest, whatever were the irregularity in the figure, or the density of the vibrating body.

Huygens, the profound and original author of the Theory of the Pendulum, had demonstrated that the centres of suspension and oscillation are convertible with one another; or that, if in any pendulum the centre of oscillation be made the centre of suspension, the time of vibration will be in both cases the same. Hence, conversely, said Captain Kater, if the same pendulum with different points of suspension can be made to vibrate in the same time, the one of these points must be the centre of oscillation when the other is the centre of suspension : and thus their distance, or the true length of the pendulum is found. It is curious to remark, that a proposition, so well known, and affording so direct a solution of the difficulty in which experimenters on this subject had always found themselves involved, was never before, at least in as much as we have been able to discover, applied to a purpose for which, now that the secret is known, it seems so excellently and so plainly adapted. But it is one of the prerogatives of true genius, to find the highest value in things which ordinary men are trampling under their feet.

To reduce the principle just mentioned into a tangible form, some further contrivance was still necessary. We copy the author's description of his convertible pendulum.

" The pendulum is formed of a bar of plate brass, one inch and a half wide, and one-eighth of an inch thick. Through this bar two triangular holes are made, at the distance of 39.4 inches from each other, to admit the knife edges that are to serve for the axes of suspension in the two opposite positions of the pendulum. Four strong knees of hammered brass, of the same width with the bar, six inches long, and three quarters of an inch thick, are firmly screwed by pairs to each end of the bar; so that when the knife edges are passed through the triangular apertures, their backs may bear steadily against the perfectly plane surface of the brass knees, which are formed as nearly as possible at right angles to the bar. The bar is cut of such a length that its ends fall short of the extremities of the knee-pieces about two inches.

" Two slips of deal, 17 inches long, are inserted at either end, in the spaces thus left between the knee-pieces unoccupied by the bar, and are firmly secured by screws. These slips of deal are only half the width of the bar; they are stained black, and a small whalebone point inserted at each end indicates the extent of the arc of vibration.

" A cylindrical weight of brass, three inches and a half in diameter, and weighing about two pounds seven ounces, has a rectangular opening in the direction of its diameter, to admit the knee-pieces of one end of the pendulum. This weight, being passed on the pendulum, is so firmly screwed in its place as to render any change impossible."

This weight, it must be observed, is not between the knife edges, but is very near to one of them.

" A second weight, of about seven ounces and a

half, is made to slide on the bar, near the knife edges, at the opposite end ; and it may be fixed at any point on the bar by two screws, with which it is furnished. A third weight, or slider, of only four ounces, is moveable along the bar, and is capable of nice adjustment, by means of a screw and a clasp. It is intended to move near the centre of the bar, and has an opening, through which may be seen divisions of twentieths of an inch engraved on the bar."

It is by means of this moveable weight that the direction of the vibrations in the two opposite positions of the pendulum are adjusted to one another; after which it is secured immoveably in its place.

The knife edges, or prisms, which make so important a part of this apparatus, and are to serve alternately as the axes of motion, are made of the steel prepared in India, and known by the name of *wootz*. The two planes which form the edge of each prism are inclined to one another nearly at an angle of 120 degrees. Every precaution was used to render the edges true, or straight, and to give the hardest temper to the steel; and a long series of experiments proves fully that they have been successful. Every precaution was also taken to give stability to the axes of suspension, when the experiments were made: But for the details of these, we find it necessary to refer to the paper itself.

We come now to the very ingenious method which Captain Kater adopted for determining the number of vibrations made by his pendulum in twenty-four hours. It is no doubt sufficiently understood, from what has been already said, that the pendulum was not to be applied to a clock, nor to receive its motion from any thing but its own weight. When experiments of this kind were attempted, it was for a long time supposed that the pendulum might safely be permitted to receive the continuance of its motion from machinery ; and that, as it was then in no danger of coming to rest, the results were more to be depended on. This conclusion, however, proceeded on a great mistake as to the part which the machinery of the clock performs on such occasions. That machinery is hardly ever, we believe, so nicely adjusted as accurately to restore to the pendulum the motion it loses in each vibration, (from friction about the centre, and from the resistance of the air,) without either allowing any defect, or producing any excess. A clock, in general, accelerates the natural motion of the pendulum, and forces it to vibrate faster than it would do if impelled only by its own gravity. In experiments, therefore, where the relation of the length of the pendulum to the time of vibration is to be determined, the clock can only be used to measure out a given portion of time, or to assist in numbering the vibrations.

The manner in which this last can be done, is not so obvious as may be imagined. The mere counting of the vibrations one by one, and marking the number at stated intervals of time, would be a very inconvenient and imperfect way of going to work. As the experiment must be long continued, and frequently resumed, the *tedium* and irksomeness of counting the vibrations would become great, and, like every labour that is tedious and irksome, must be in danger of being inaccurately performed, more especially by *mathematicians*, the persons into whose hands the operation is most likely to fall. Even if no error were committed, there would still be an insecurity which nothing could remove. It is, indeed, the business of every experimenter to throw as great a share of the responsibility as he can on his apparatus, or on the physical agents he employs : and as little as possible on himself and his *living* assistants. Different means have accordingly been used for avoiding the above inconveniences ; and of those that we are acquainted with, we think Captain Kater's is the best, the least tedious, and the most infallible.

Boscovich, in the 5th volume of his Opera Opt. et Astr. gives an account of a method which he had employed, and which he ascribes to Mairan.

A clock being well regulated, according to mean time, and having its case open, the experimental pendulum was placed right before it at a little dis-

tance, with its point of suspension firmly supported.
The position of both was such, that, in their state
of rest, the pendulums were seen by a person
placed in front of them, coinciding with one another, and with a vertical line drawn on the clock-
case behind the pendulum. That this coincidence
might be more distinctly seen, when it happened
to the moving bodies, it was viewed through a hole
in a piece of paper fixed to the back of a chair on
the opposite side of the room. The two pendulums having been put in motion, and not vibrating
exactly in the same time, one would gain upon
the other, and after a while they would be seen
through the hole in the paper to coincide with one
another, and with the fixed line on the body of the
clock. The instant of this coincidence must be
noted. When they next coincide, the difference
of the times of their vibrations must have amounted to one entire vibration. This is also to be
noted; and thus the information of the clock will
give the ratio of the time of its own vibrations to
the time of those of the pendulum. This experiment must be often repeated, and a mean taken,
that if there are any accidental errors, there may
be a probability of their balancing one another.

The method of numbering the vibrations in the
experiments of Borda and Cassini, was similar, in
many respects, to the preceding, and may have

been suggested by the same to which Boscovich refers, that of their ingenious countryman Mairan.

The pendulum was placed, as in the former example, right before the clock with which it was to be
compared, so that the wire by which the platina
ball was suspended, bisected the ball of the clock
pendulum when at rest; the middle point of this
last being marked by the intersection of two white
lines drawn on a black ground.   The two pendulums were viewed through a small telescope, fixed
on a stand on the opposite side of the room, and a
screen was also placed before the pendulums, the
edge of which just covered the wire of the platina
pendulum, and therefore concealed behind it one
half of each of the balls.   The platina pendulum
was nearly 12 feet long; so that it made about one
vibration while the pendulum of the clock made
two.

Suppose, now, that when the pendulums were
put in motion, the wire disappeared behind the
screen, before the cross; as the times of the vibrations are not supposed accurately as 2 to 1, it
would happen that the interval between the disap-
pearances would decrease, till at length both objects came to pass behind the screen at the same
instant.   The instant of this first coincidence was
observed; the oscillations then began to disagree,
afterwards to approach, till at length a second coincidence took place.   In the interval between the

coincidences, the clock had gained two seconds on the pendulum ; so that the ratio of the times of the vibrations of the two pendulums was given. *

Captain Kater's pendulum was compared with two clocks, the property of H. Browne, Esq., in whose house the experiments were made. One of these, a time-piece by Cumming, is of such excellence, that the greatest variation of its daily rate, from the 22d of February to the 31st of July, did not exceed three-tenths of a second. The clock, however, with which the immediate comparison was made, and in front of which the pendulum was placed, was one of Arnold's, also of excellent construction. The pendulum was securely suspended in front of this last, and close to it, so that it appeared to pass over the centre of the dial-plate, with its extremity reaching a little below the ball of the pendulum. A circular white disk was painted on a piece of black paper, which was attached to the ball of the pendulum of the clock, and was of such a size, that, when all was at rest, it was just hid from an observer on the opposite side of the room, by one of the slips of deal which form the extremities of the brass pendulum. On the opposite side of the room was fixed a wooden stand, as high as the ball of the pendulum of the clock, serving to support a small telescope, magnifying about four times. A diaphragm in the focus

---

* Bàse du Syst. Métrique, Tom. III. p. 343.

was so adjusted as exactly to take in the white disk, and the diameter of the slip of deal which covered it.

" Supposing now both pendulums set in motion, the brass pendulum a little preceding the clock, the slip of deal will first pass through the field of view at each vibration, and will be followed by the white disk. But the brass pendulum being rather the longer, the pendulum of the clock will gain upon it ; the white disk will gradually approach the slip of deal, and at length, at a certain vibration, will be wholly concealed by it. The instant of this total disappearance must be noted. The pendulums will now appear to separate ; and, after a certain time, will again approach each other, when the same phenomenon will take place. The interval between the two coincidences will give the number of vibrations made by the pendulum of the clock ; the number of vibrations of the brass pendulum is greater by two."

Thus was determined the number of vibrations made by the brass pendulum in a given interval of time ; and so, by proportion, the number for a whole day. The interval between the two nearest coincidences was about $132\frac{1}{2}''$; and four of these, that is, five successive coincidences, gave an interval of $530''$ or 8 minutes 50 seconds ; after which, the arc described by the brass pendulum became too small. The pendulum was then stopped, and

put in motion anew as oft as it was judged proper
to repeat the observations.

Being now in possession of the means of determining, with great accuracy, the number of vibrations performed by his pendulum in a given time,
Captain Kater proceeded, by reversing it, to make
the vibrations equal in its two opposite positions.
The sliding weight mentioned above was used for
producing this equality; which, after a series of
most accurate and careful experiments, was brought
about with a degree of precision that could hardly
have been anticipated. By the mean of 12 sets of
experiments, each consisting of a great number of
individual trials, with the end of the pendulum
which we shall call A, uppermost, the number of
vibrations in twenty-four hours was 86058.71;
and, with the same end A, lowest, the mean of as
many others gave 86058.72, differing from the
former only by a hundredth part of a vibration.
The greatest difference was .43, or less than a half.
Such exactness, we believe, has never been exceeded; and would hardly be thought possible, if the
data from which so satisfactory a result was deduced were not given in full detail in the paper before
us.

Thus, for the first time, after having been an oc-
casional object of research for more than 150 years,
has the centre of oscillation of a compound pendulum been found by experiment alone, according to

a method also of universal application, and admitting of mathematical precision. The ingenious author has therefore the honour of giving the first solution of a problem, extremely curious and interesting in itself, independently of its immediate connection with one of the greatest and most important questions in the natural history of the Earth.

The next thing to be done, was to measure the length of the pendulum, or the distance between the knife edges, which had alternately served as the centres of suspension and oscillation, and from thence to deduce the length of the pendulum vibrating seconds in the latitude of London, which, at the spot (Mr Browne's house in Portland Place) where the observations were made, is 51° 31′ 8″.4. It is sufficient here to state, that no expedient has been neglected that practical or theoretical science is at present in possession of, for giving precision to this measurement, and that it was in all respects such as to correspond to the accuracy of which we have just seen so striking an example. Including the effects of temperature, of the buoyancy of the atmosphere, of the shortening of the arcs of vibration from the beginning to the end of each trial, and reducing the actual vibrations to those in arcs infinitely small, the length of the seconds pendulum from a mean of the 12 sets of experiments above mentioned, comes out 39.13829 inches, or 39.1386, reducing it to the level of the sea. * The greatest difference between this result and any one of the 12 of which it is a mean, is .00028 of an inch ; that is, less than three of the ten thousandth parts. The mean difference among these results, adding the positive and negative together, as if they had all one sign, or were all on the same side, is little more than one ten thousandth of an inch ; and as the above is obviously a supposition more unfavourable than ought to be made, we think the probability is very great that the preceding result does not

---

* The scale on which this pendulum is measured, is Sir George Shuckburgh's, the work of Troughton, and of the highest authority. It is described by Sir George in the Phil. Trans. for 1798. Gen. Roy's scale, which is very important, as being that from which are derived all the measurements in the trigonometric survey, was compared with the preceding by Captain Kater. So also was the yard on what is called the parliamentary standard, which was laid off by Bird, but it would seem not so carefully as might have been expected. The scales in the order in which they are now named, appear from these measures to be as the numbers 1 ; .99963464 ; 1.00000444.

In another communication from Captain Kater, in the same volume of the Phil. Trans. the length of the French mètre is compared with the yard on Sir G. Shuckburgh's scale. He found the *mètre* as marked by two very fine lines on a bar of platina=39.37076 inches on his scale ; as marked by the ends of a metal rod in the usual way, the *mètre*= 39.37081. Supposing the two of equal authority, the mean length of the mètre is 39.37074 inches. The temp. of the scale 62° of Fahr.

err so much as a unit in the last decimal place, or
in that which denotes ten thousandths of an inch.

The determination given above is considerably
different from that which had been received on the
authority of the older experiments The length
given to the seconds pendulum, in the bill for the
equalization of weights and measures, is 39.13047,
differing from that just assigned by .00813; a considerable quantity, in a matter where it appears that
a ten thousandth of an inch is a distinguishable
magnitude.

To the paper which ends with the *measures* just
given, is added, in an appendix, a letter from Dr
Thomas Young, containing a demonstration of a
very remarkable property of the pendulum recently discovered by M. Laplace. The property is, that if the supports of a pendulum, inverted as above described, be two cylindric surfaces,
the length of the pendulum is truly measured by
the distance of those surfaces. This applies immediately to the experiments we have been considering; because the knife edges, supposing them
somewhat blunted, may be regarded as cylindric
surfaces of very great curvature, or of very small
diameter; and in this way, as Dr Young very
justly remarks, is removed the only doubt that can
reasonably be entertained of the extreme accuracy
of the conclusions. The theory of experiments
made with the inverted pendulum is therefore

much indebted to the calculus of the profound mathematician above named. We have not seen his analysis ; but a demonstration is sketched by Dr Young, that seems sufficiently concise and simple, considering the recondite nature of the truth to be demonstrated.

Captain Kater's paper is dated in July 1817, the experiments described in it having been made previously to that time. The same apparatus that was thus perfected has been employed since, for the purpose of ascertaining the length of the seconds pendulum in different latitudes, with a view to the questions about the figure of the earth. That the precise object of the experiments may be the better understood, it may be proper to go back to the summer 1816.

After the bill for the equalization of *weights* and *measures* was thrown out, the attention of those who promoted the scheme of equalization, was naturally turned to the determination of the lengths of the pendulum ; so that one of the good effects arising from the disappointment of the premature plan of equalization, was probably that of directing the ingenuity of the author of this paper to a subject in which it has been so successfully exerted. This other good effect also resulted from it. The French academicians were known to have directed a great deal of attention to this subject ; the experiments of Borda and Cassini, so often mentioned, were the most accurate that had yet been made ; and the speculations of Laplace had deduced, from a collection of the best experiments that he could find, some very important conclusions concerning the figure of the earth.   On this subject, however, more information was still to be expected, when experiments of equal accuracy with those made at Paris should be repeated in different latitudes.   It would then be seen, whether the lengths of the pendulum agreed in giving the same figure to the earth with the measures of degrees of the meridian, and, if they did not, in what respects they differ- ed.   This was the more desirable, that some in- consistencies had been found in the information derived from the last of these sources, and that there was reason to think that the same causes of inconsistency might not affect the experiments made with the pendulum.   The pendulum mea- sures the intensity of gravity ; but its vibrations are little affected by the direction of that force. The measures of degrees, on the other hand, are extremely sensible to whatever affects the direc- tion of gravity, but not much to what only changes its intensity.   Hence, each of these methods of inquiring into the figure of the earth contains a remedy for the imperfections of the other ; each by itself is incomplete ; and both, of course, ought to be employed.

It has been imagined, that the intensity of gravity suffers less alteration from the action of local causes than the direction does ; and that, on that account, the conclusions deduced from the pendulum are more likely to be free from inconsistency than those that depend on the measurement of degrees. But it must not be supposed that, with the pendulum carried to its present state of sensibility and precision, the results will be free from inconsistency, or beyond the influence of the local irregularities that may exist immediately under the surface of the earth. Were the pendulum the same inaccurate instrument that it was a few years ago, it might not feel the influence of such causes as only increase or diminish the intensity of gravity by a very small part of the whole. But, when the length of the pendulum can be determined to the ten thousandth of an inch, or to $\dfrac{1}{134959}$ of its whole length, the force of gravity is measured with the same precision, and one part out of 134959 is rendered sensible. Now, it seems to us probable, that the variation in the density of the strata immediately under the surface, may produce a change in the intensity of gravitation, much more considerable than one part in 134959 ; the pendulum will not fail to be affected by this irregularity, and to give information of it. The force with which Schehalien disturbed the plumb-line was about $\dfrac{1}{34376}$ of gravity, or nearly four parts in 134959.

We think that, without any exaggerated suppositions, by the presence of an extensive stratum of gneiss, or of hornblende schistus, or of any great body of granite immediately under the surface at one place, and of chalk, common sandstone or limestone at another, a difference in the intensity of gravity, even greater than the preceding, may be readily produced. The extreme sensibility to which the apparatus of the pendulum has been brought by Captain Kater, though it adds infinitely to the value of the instrument, will not, probably, add to the consistency of its reports. On that very account, however, those reports will afford more important information concerning the constitution of the Globe; and the manner of extracting from them the most probable average result is also sufficiently understood.

We venture to throw out these conjectures before the new results have been communicated, (except those of Paris and London;) and if we are wrong, we have the satisfaction to know, that our error will be soon corrected.

As the Academy of Sciences was already engaged in experiments of the same kind with those which were to be undertaken under the direction of the Royal Society of London, it was resolved by the latter, on the motion, we believe, of the President, to invite the former to authorize some of its members to join in the experimental and astronomical researches of which England was about to become the theatre. The invitation was accepted; the Governments of both countries signified their acqui-escence, and offered their support; and the friends of science every where rejoiced in this mark of cor-diality exchanged between two societies which the misfortunes of Europe had so long placed at a dis-tance from one another. In the beginning of the summer 1817, M. Biot arrived in England, fur-nished with an apparatus for determining the length of the pendulum, the same, we believe nearly, that was used by Borda and Cassini. It was agreed that observations on the length of the pendulum should be made at London, at Edinburgh, and at the northern extremity of the greatest arc of the meri-dian that was to be determined by the trigonome-trical survey of Britain, which, as was already known, must terminate in Shetland, between the small islands of Unst and Balta. M. Biot, accom-panied by Colonel Mudge, his son Captain Mudge, and Dr Olinthus Gregory, repaired to Edinburgh, and, having made observations at Leith Fort, em-barked for Shetland. They were joined by Captain Colby, who conducted the trigonometrical survey, and who, with the zenith sector, was about to ob-serve the highest latitude to which his system of triangles would extend. Colonel Mudge was forced, by bad health, to return; M. Biot and Dr Gre-gory made their observations separately, but in the same small island; and the former conti-nued till late in the season on the barren rock,

where he was almost left alone, surrounded by a
stormy sea, and a dusky and inclement sky.   The
spirits of a man accustomed to the finer climates of
the south, must have sunk in such a situation, had
they not been supported by his love of science, and
his zeal for promoting its interests.   He has written an account of his visit to Great Britain, and
particularly of his reception in Scotland and the
Isles, drawn up in an excellent spirit, full of good
temper, cheerfulness, and a disposition to be pleased ; and abounding also in judicious remarks.   The
Shetland Isles seem particularly to have interested
him ; and the contrast between the aspects which
the moral and physical world presented in that remote region, to have struck him forcibly.   He was
pleased with the kindness, hospitality, and intelligence of his hosts ; and they, no doubt, were filled
with respect for an illustrious stranger, who, from
the centre of civilization, had penetrated into their
distant isle, and was connecting, with the researches
and the renown of Science, the obscure and se-
questered corner in which Providence had fixed
their habitation.   He must have experienced feelings of high gratification, on considering that he
had now assisted in defining both extremities of a
line, extending from the most southerly of the
Balearic to the most northerly of the Shetland
Isles, the longest that the finger of Geometry had
yet attempted to trace, or her rod to measure, on
the surface of the earth ;—a work that, in all ages,

it will be the boast of the nineteenth century to have accomplished.   The different aspects of nature, at the remote stations, which he had successively occupied, would not fail to present themselves with all the force that contrast can bestow ;—the bright sun, the cloudless skies of the south, the glowing tints and the fine colouring of the Mediterranean, compared with the misty isle on which he now stood, and the tempestuous ocean which was raging at his feet.   If he turned to the moral world, the contrast was also great, but it was reversed ; and he would, perhaps, think of the fierce barbarians before whom he or his companions had been forced to fly, when the lonely islander was opening his cottage to receive him, and defend him from the storm. He would not then fail to reflect, how much more powerful moral causes are, than physical, in determining the good or evil of the human character.

M. Biot, on his return to London in the autumn, was joined by MM. Arago and Humboldt, and, in conjunction with these illustrious associates, completed his experiments.   The results have not yet, we believe, been given to the public ; neither have those of Dr Gregory.   The scientific world waits impatiently for both.

During the present summer Captain Kater has visited the same stations, as well as some others particularly connected with the trigonometrical survey, employing the apparatus above described for ascertaining the length of the pendulum.   The result of observations made at six different points, from Unst in Shetland to Dunnose in the Isle of Wight, may be expected in the course of the ensuing winter. A great advantage that results from the manner in which his experiments are made, is the comparative shortness of the time that they take up. After the rate of the clock has been ascertained, the observations of the pendulum may be finished in three or four days, and the number of its vibrations in twenty-four hours, determined within a fraction of a second. Thence the length of the seconds pendulum is easily deduced, being, to that of the invariable pendulum used in the experiment, and of which the length is already accurately known, as the square of the number of vibrations performed by this last in twenty-four hours, to the square of 86,400, the number of seconds in the same time. When the experiments are conducted in the way followed by the French astronomers, the length of the pendulum must be measured anew at every station. We cannot help thinking, that the frequent repetition of an operation, which it is always difficult to perform with accuracy, ought as much as possible to be avoided.

While we are concluding this article, we learn, with great satisfaction, the farther progress of other operations connected with those of which we have been giving an account. Captain Colby, after finishing his campaign among the Scottish mountains, is at this moment on his way to Dunkirk, for the purpose, as we suppose, of joining the French mathematicians, in order to examine, over again, the junction of the English and French triangles, and to determine the latitude of the extreme point of the meridian of Paris with the zenith sector—the same excellent instrument that has been used for all the celestial observations in the British survey. As this will involve a comparison between that sector and Ramsden's great theodolite on the one hand, and the repeating circle on the other, it will be an experiment of great interest to astronomers; and, we believe, the conduct of it could not be in better hands than those into which it is about to be committed. Orders, we understand, have been given by Lord Liverpool for preparing every thing that may be required along the coast of Britain. The liberality and steadiness with which Administration has supported the trigonometrical survey from its commencement, is deserving of the greatest praise, and is a strong claim to the gratitude of the Scientific World.

FINIS.

Printed by George Ramsay & Co.
Edinburgh, 1821.

Printed in the United States
By Bookmasters